Oil and Gas in Africa

Oil and Gas in Africa

Joint Study by the African Development Bank and the African Union

OXFORD
UNIVERSITY PRESS

OXFORD
UNIVERSITY PRESS

Great Clarendon Street, Oxford OX2 6DP

Oxford University Press is a department of the University of Oxford.
It furthers the University's objective of excellence in research, scholarship, and education by publishing worldwide in

Oxford New York

Auckland Cape Town Dar es Salaam Hong Kong Karachi
Kuala Lumpur Madrid Melbourne Mexico City Nairobi
New Delhi Shanghai Taipei Toronto

With offices in

Argentina Austria Brazil Chile Czech Republic France Greece
Guatemala Hungary Italy Japan Poland Portugal Singapore
South Korea Switzerland Thailand Turkey Ukraine Vietnam

Oxford is a registered trade mark of Oxford University Press in the UK and in certain other countries

Published in the United States
by Oxford University Press Inc., New York

First published 2009

British Library Cataloguing in Publication Data

Data available

Library of Congress Cataloging in Publication Data

Data available

Typeset by SPI Publisher Services, Pondicherry, India
Printed in Great Britain
on acid-free paper by
CPI Antony Rowe, Chippenham, Wiltshire

ISBN 978–0–19–956578–8

1 3 5 7 9 10 8 6 4 2

Foreword

This report is jointly produced by the African Development Bank (AfDB) and the African Union (AU). It analyzes the status of Africa's considerable oil and gas resources and highlights—with an African focus but a global perspective—the challenges and constraints facing the continent in the exploitation and utilization of these resources. The report has a dual perspective: (1) The resources can be used strategically to accelerate growth in both net oil-exporting and net oil-importing African countries; and (2) A framework for a coherent future vision will help African governments and oil and gas industry operators in their decision-making. The report provides updated and highly valuable data and analysis. It analyzes the roles of the AfDB and the AU, considering their different but complementary mandates geared towards the development of the continent. The report includes recommendations on future directions and options for maximizing the benefits of Africa's oil and gas resources.

The key challenge for harnessing oil and gas resources is making the right strategic choices and synchronizing their implementation in a context that supports fiscal prudence and minimizes macroeconomic distortions. This should be backed by adequate institutional capacity and national and local level participation in oil and gas revenue management. Given that oil is a non-renewable resource, it is vital to reduce corruption, promote intergenerational equity, and negotiate more beneficial and transparent contracts with oil companies.

Regional cooperation is an important element in addressing the development challenges resulting from changes in international oil markets. Such cooperation can cover many areas, for example, infrastructure development, but efforts to channel oil revenues so that they can be used to meet the continent's development needs are particularly important. Consequently, this report also outlines prospects for establishing an African Petroleum Fund.

The African Development Bank, in cooperation with the African Union and other partners, will scale up its efforts and continue to work with African governments to design and implement strategies for optimal utilization of oil and gas resources to accelerate growth and development in the continent. In particular, the Bank is committed to helping countries use oil and gas resources to improve the lives of the poor through increased investment in health, agriculture, education, physical infrastructure, and other non-oil sectors.

Through better management and allocation of oil and gas revenues, countries can promote economic diversification and thus build the basis for higher and stable growth. This report provides valuable input toward this goal.

Louis Kasekende
Chief Economist
African Development Bank

Elham M. A. Ibrahim
Commissioner for Infrastructure and Energy, African Union

Acknowledgements

This Report was produced under the overall guidance of Louis Kasekende, Chief Economist of the African Development Bank and Bernard Zoba, former AUC Commissioner for Infrastructure and Energy. Direct supervision came from Temitope W. Oshikoya and Léonce Ndikumana, former and current Directors, respectively, of the Development Research Department of the African Development Bank.

The research team that worked on the study and on the resulting report comprised Abdul B. Kamara (Manager of the Research Division), John C. Anyanwu (Task Manager for the study), Michael Juel (Technical Assistant), Salvador Mondlane (Consultant), and Akin Iwayemi (Consultant).

The following African Development Bank and African Union Commission staff provided critical inputs, valuable contributions and support at various stages of the study: Désiré Vencatachellum, Rebei Noorman, Rasheed Oyaromade, Septime Martin, Hee-Sik Kim, Charles Lufumpa, Beejay Kokil, Louis Kouakou, Ibrahim Diallo, Nirina Letsara, and Mbiya Kadisha from the African

Development Bank; and Aboubakari Baba-Moussa, Philippe Niyongabo, Bernard Barandereka, and Hussein Elhag from the African Union Commission.

Rick van der Ploeg of Oxford University provided valuable comments and suggestions on the background analytical work and reviewed earlier drafts of the manuscript.

The Report team appreciates competent editorial work by Felicia Avwontom as well as administrative assistance by Rhoda Bangurah, Abiana Nelson, and Ines Hajri.

Contents

List of Figures, Tables, Boxes, and Map

List of Figures

List of Tables

List of Boxes

Map

Acronyms and Abbreviations

AfDB/ADB	African Development Bank
ADR	African Development Report
AFD	Agence Française de Développement
APF	African Petroleum Fund
APRM	African Peer Review Mechanism
AU	African Union
AUC	African Union Commission
B/D (bpd)	Barrels per day (crude oil)
BCM	Billion cubic meters
BTU	British Thermal Units
CGE	Computable General Equilibrium
CIF	Cost Insurance and Freight
CNG	Compressed Natural Gas
COMESA	Common Market for Eastern and Southern Africa
DFID	Department for International Development

ECOWAS	Economic Community of West African States
EIA	Energy Information Administration
EITI	Extractive Industries Transparency Initiative
EPFI	Equator Principles Financial Institutions
ETBE	Ethyl Tertiary Butyl Ether
EU	European Union
FDI	Foreign Direct Investment
GDP	Gross Domestic Product
HDI	Human Development Index
ICMM	International Council on Mining and Metals
IFC	International Financial Corporation
IISD	International Institute for Sustainable Development
IMF	International Monetary Fund
JV	Joint Ventures
KPCS	The Kimberley Process Certification Scheme
Ktoe	Thousand (kilogram) tons of oil equivalent
LNG	Liquefied Natural Gas
MDGs	Millennium Development Goals
MMSD	Mining, Minerals and Sustainable Development (project)
MNOC	Multinational Oil Companies
NEITI	Nigerian Extractive Industries Transparency Initiative

NEPAD	New Partnership for Africa's Development
NGO	Non-Governmental Organization
NNPC	Nigerian National Petroleum Corporation
NOEC	Net Oil-Exporting Country
NOIC	Net Oil-Importing Country
NYMEX	New York Mercantile Exchange
O&G	Oil and Gas
ODA	Official Development Assistance
OECD	Organization for Economic Cooperation and Development
OPEC	Organization of the Petroleum Exporting Countries
PFM	Public Financial Management
PRSP	Poverty Reduction Strategy Paper
PSA	Production Sharing Arrangement
PWYP	Publish What You Pay (campaign)
RMC	Regional Member Countries
SADC	Southern African Development Community
SC	Service Contract
SIDA	Swedish International Development Agency
SOC	State Oil Companies
UK	United Kingdom

UNDP	United Nations Development Program
UNECA (ECA)	United Nations Economic Commission for Africa
UNEP	United Nations Environmental Program
U.S. (USA)	United States (of America)
USAID	United States Agency for International Development
USD	United States Dollar
WB	The World Bank

Executive Summary

Africa has considerable oil and gas resources that can help accelerate growth on the continent if used strategically. Although new resources are discovered progressively, they are not equally distributed; indeed, 38 out of 53 African countries are currently net oil importers. High and volatile oil prices are thus a challenge for all of Africa; they represent an opportunity to be pursued for exporting countries and an obstacle to be tackled for importing countries.

The broad objective of this report is to shed light on key issues related to the social, environmental, and economic impacts of high and volatile oil and gas prices. This includes a discussion of Africa's oil and gas status and its major challenges and opportunities in the energy sector.

Global Perspective and Status of Oil and Gas Resources in Africa

Energy is an indispensable input for economic growth and social development. Two-thirds of global energy requirements are met with oil and gas supplies. Conventional wisdom holds that energy consumption per capita is strongly correlated with the level of economic and social progress. Remarkably, the three non-renewable fossil fuels, oil, natural gas, and coal, constitute almost 90 percent of commercial energy consumed globally.

The regional composition of global energy consumption reveals a wide disparity in global use and access to commercial energy. Although Africa has about 15 percent of the world's population, it consumes only 3 percent of global commercial energy. The paradox is that Africa's share in global energy production is about 12 percent, and trending upwards.

The evolution of world energy markets in the post-1970 period has been dramatic and its impact on the world economy and politics profound. This is illustrated by the worldwide economic ripple effects caused by price volatility and occasional spectacular spikes in the prices of the dominant global energy resources—oil and gas. World oil prices have trended ever higher since 2000, and natural gas prices have tracked along. Some of the reasons for the rise in oil prices include rising demand in emerging economies, especially in China and India, declining spare capacity in major producing countries, peaking of production in several important oil-producing areas, and lack of expansion in refinery capacity.

Driven by continued high demand in the Western world, coupled with significant and accelerating new demand from emerging economies, such as China, India, and Brazil, global energy consumption is expected to grow by more than 50 percent in the first quarter of this century. Oil and natural gas are expected to be in particularly high demand by 2025, with global oil consumption (demand) projected to rise by 57 percent. It is very unlikely, even taking into account the massive investments in the energy sector around the world, that the oil and gas industry will be able to produce and deliver sufficient energy to meet global demand. By some projections, the "peak oil" production has already been reached, or will be reached in a few years. The ensuing shortages, coupled with concomitant rising energy prices, will place significant pressure on net oil-importing societies in Africa if not addressed strategically and aggressively.

Africa is endowed with vast quantities of both fossil and renewable energy resources. Furthermore, it is the main continent in the world with frequent and substantial new findings of oil and gas: In the past 20 years, oil reserves in Africa grew by over 25 percent, while gas grew by over 100 percent. Africa's rich oil fields and the prospects for more discoveries have transformed it into an important player and a key "target" in global oil production and resource extraction. Oil production in the continent is expected to continue to rise at an average rate of 6 percent per year for the foreseeable future. The majority of oil reserves (and production) in Africa comes from Libya, Nigeria, Algeria, Angola, and Sudan, which together produce more than 90 percent of the continent's reserves. Proved natural gas reserves in Africa are mainly concentrated in four countries—Algeria, Egypt, Libya, and Nigeria—which possess 91.5 percent of the continent's proved reserves. In particular, Nigeria's undeveloped natural gas reserves are a logical target of the international giants in the sector. Furthermore, large deposits of natural gas have been identified in Tanzania; significant oil deposits found in Albertine Graben in Uganda and in the western part of Ghana (offshore); and there are potential significant oil discoveries in South Africa, Mozambique, and Tanzania.

Considering the current uncertainties about energy supply, the key drivers of future demand, the policies of consumer countries (especially with respect to nuclear and other alternatives to oil and gas), and expected future global economic growth and technology development, there is need to clearly establish Africa's position and develop strategies for future supply adequacy. The energy crisis due to high oil prices, the environmental impacts of oil production, and the growing concern about the viability of oil-based fuels and products are leading to initiatives to find and develop alternative energy sources.

Maximizing the Benefits from Oil and Gas Resources

A key concern regarding the governance of oil and gas resources is that the governments of African oil- and gas-producing countries receive an inadequate share of the large rents from production. This may stem from a number of reasons, including contracts and regimes that are not designed to extract maximum rents; and oil and gas policies that are designed primarily to promote and attract investments and have not evolved with changing global dynamics and national interests. The sustainable development of oil and gas resources requires policies, principles, and practices that support the utilization of resources in a manner that does not prevent future generations from benefiting from the resources. A great challenge, particularly for oil-producing African countries, is to ensure sufficient, reliable, and environmentally responsible supplies of oil, at prices that reflect market fundamentals. To achieve this important goal, several challenges have to be addressed, including high and volatile oil prices; growing external and internal demand for oil; increasing import dependence of many African countries; and, most importantly, sustainable management of the continent's oil and gas resources for the benefit of all. The regional nature of these challenges and the growing interdependence between net importing and net exporting African countries require a strengthened partnership among all stakeholders to enhance regional energy security.

The sustainable management of oil and gas also faces the challenge that large natural resource revenues tend to replace more stable and sustainable revenue streams, exacerbating existing problems related to development, transparency, and accountability. This tends to free natural resource-exporting governments from the types of citizen demands for fiscal transparency, and accountability that arise when people pay taxes directly to the government. Thus, natural resource export earnings actually sever important links between the people and their governments—links that are related to popular interests and control mechanisms. Governance indicators such as government effectiveness, voice and accountability, political instability and violence, the rule of law, regulatory quality, and control of corruption are correspondingly markedly weaker in oil-rich African countries.

Despite the challenges and issues involved, an oil and gas resource boom can, under the right circumstances, be an important catalyst for growth and development. The often-referred-to "natural resource curse" can be avoided with the right institutions and policies. Several countries in Africa have demonstrated this and there is some reason for cautious optimism that more countries have learned hard lessons from past resource booms, and, in future, will pursue strategies and policies that will allow them to fully reap the benefits of their natural resource wealth.

Impact of Rising Oil Prices

The AfDB has developed a new open-economy macroeconomic model to quantify the impact of high oil prices on oil-importing and oil-exporting African economies. The results of the model estimations suggest that high oil prices can have very harmful effects on the economies of African oil-importing countries, especially those that are heavily debt-burdened. The rising oil prices will lead to a decrease in output and consumption, and to a worsening of the net foreign asset position. For median oil-importing countries, the five-year cumulative output loss resulting from a doubling in the price of oil can be as large as 23 percent under a fixed exchange rate regime. However, this recessionary effect can be cushioned through government intervention in the form of limited pass-through of the oil price increase or through foreign aid.

Coping with high oil prices calls for a set of measures to maximize the positive impacts and mitigate the negative ones. Specific measures (approaches) that governments can implement include:

- Price-based policies (comprising full passing on of oil price increases and granting of targeted subsidies to some vulnerable end-users);
- Policies to reduce the supply costs (hedging, economies of scale, security of stocks);
- Implementing quantity-based policies (reduce demand, increase supply); and
- Achieving diversification into non-petroleum sources of energy (from oil to natural gas; and more generally from fossil fuels to renewable power sources).

Coping mechanisms depend on the government to formulate policies that can work sustainably for pro-social development and secure stability, for instance, by applying cross-subsidies to specific vulnerable groups. Coping policy mechanisms can also include management of inventories, that is, securing an adequate strategic reserve as a defensive stockpile.

Challenges, Opportunities, and Policy Issues

Oil and gas resources are a principal source of public revenues and national wealth for the governments of net oil-exporting countries. As a major source of wealth and energy in Africa, oil and gas resources are critical for growth, development, and good governance. They also pose major management challenges for African governments. The issues go beyond technical management of oil and gas resources and collection of revenues. Resource control, governance, transparency in the utilization of resource wealth for development, preserving and optimizing the resource base, environmental protection, and securing equitable and intergenerational long-term benefits (especially poverty reduction) are among the many critical ingredients that should be embedded in any coherent strategy aimed at harnessing oil and gas resource wealth. In other words, achieving sustained growth from oil and gas resources entails managing and enhancing the status of a complex portfolio of natural, human, and social capital.

Recommendations

The recommendations arising out of the analysis and discussion in this report cover the continental, regional, and country levels, distinguishing between oil importers and exporters.

Continental Level

At this level, it is important for continental bodies, mainly the AU and the AfDB, to become more engaged in helping African countries manage their oil and gas resources. It is likewise very important that these continental bodies, together with other international donors and stakeholders, provide technical and financial assistance to help new oil and gas producers negotiate fair and pro-development contracts where environmental, social, and revenue distribution are an integral part of the management of oil and gas.

Regional Level

The regional economic blocs can likewise play a key supporting and coordinating role and can be instrumental in advancing the following initiatives:

1. Promote regional integration in oil and gas exploitation.
2. Build regional infrastructure (such as oil and gas pipelines) for sustainable exploitation of oil and gas.

3. Promote regional economies of scale in the sector, focusing especially on downstream industries (such as refineries).
4. Encouraging countries to adhere to principles of good governance and transparency initiatives for revenue management.
5. Promote regional sharing of experiences.
6. Promote intra-regional trade in the oil and gas sector.
7. Support and promote the African Petroleum Fund.
8. Establish regional mechanisms for sharing experiences on oil and gas issues, especially those related to contract negotiations and technology transfer.
9. Encourage the sharing of regional expertise in the sector.

Country Level: Net Oil-Producing Countries

To maximize benefits from oil resources, net oil-exporting countries should

1. Promote knowledge of verified and potential oil and gas resources (such as geological data and resource estimation), for example, establish oil and gas information database.
2. Manage oil and gas resources sustainably and promote high standards of environmental protection, including rehabilitation.

3. Secure optimal rents (royalties, taxes, shares, and other revenues) based on sound, fair negotiations (including through auctions) and "best practices" for contracts.
4. Promote good governance, accountability, and transparency in oil and gas revenue management. This also entails engaging in initiatives that can reduce corruption, such as the EITI and APRM, notably through more civil society involvement.
5. Promote revenue distribution that reaches the local and poor people in the oil fields; make revenue shares public to avoid speculations; and pursue pro-poor policies (equitable distribution of benefits).
6. Engage in social policies that promote harmony and well-being in the communities affected by oil and gas production.
7. Promote economic diversification to reduce commodity dependence (investment in the future).
8. Promote appropriate policies, notably fiscal regimes, which prevent market distortions—for example, the "Dutch disease."
9. Strengthen human, technical, and institutional capacity.
10. Design adequate investment policies to attract investors, but also to protect the resources and the interests of the country in the long term.

11. Establish mechanisms to save the excess income for future generations (intergenerational benefits), for example, by establishing sovereign wealth funds or other savings and investment funds.
12. Seek advice from regional and continental stakeholders and supporters (such as the AfDB and the AU) whenever relevant.
13. Support regional and continental initiatives that seek economic, social, and political improvements of oil, gas and energy-related initiatives, notably the African Petroleum Fund initiative.
14. Support regional infrastructure for the sustainable exploitation of oil and gas.
15. Support regional economies of scale in the downstream sub-sector.
16. Support intra-regional trade of oil and gas.
17. Share countries' experiences with regional and continental stakeholders.

Country Level: Net Oil-Importing Countries

To minimize the impact of high oil prices on their economies, net oil-importing countries should

1. Identify and recognize the direct and indirect impacts of high oil prices.
2. Develop and use alternative fuel sources such as biodiesel.

3. Find adequate mechanisms for mitigating/ smoothening the effects of high oil prices.
4. Support the African Petroleum Fund, and, when it is fully established, subscribe and contribute to its operation.
5. Identify potential downstream industries where each country has a competitive advantage (for example, a refinery in an accessible location).

Finally, the report concludes that sustainable development is about a better world for all citizens, underpinned by advances in skills, knowledge, capability, and choice. Oil and gas wealth must be used to achieve higher income per capita, better education, better health, higher life expectancy, full employment, and social stability. The AfDB and the AU have complementary strategic roles to play in pursuing and supporting Africa's energy and development future.

1

Introduction

Africa has considerable oil and gas resources that can help accelerate growth on the continent if used strategically. Although new resources are discovered progressively, they are *not* equally distributed; indeed, 38 African countries are currently net oil importers. High and volatile oil prices are thus a challenge for all of Africa; they represent an opportunity to be pursued for exporting countries and an obstacle to be tackled for importing countries.

The broad objective of this report is to shed light on the key issues related to the social, environmental, and economic impacts of high and volatile oil and gas prices. This includes a discussion of Africa's oil and gas status, from a worldwide perspective, and the continent's major challenges and opportunities in the energy sector.

1.1 Renewed Focus on Oil and Gas

For many, the twenty-first century is the "hydrocarbon century" as the dynamics of global energy markets have become distinctly marked by sharp increases in global demand and severe supply shocks that are hitting global economies. These trends are causes of concern as they affect economic performance, especially in oil-importing African countries.

Key reasons for the renewed attention on global oil shocks relate primarily to their impact on macroeconomic variables, notably inflation, the exchange rate, current account and fiscal balances. The effect of oil price shocks is more pronounced in developing net oil-importing countries, 38 of which are located in Africa (African Development Bank, 2008). These countries have been more severely hit than developed countries, partly because they are more dependent on imported oil, but mostly because energy use is far less efficient in these countries than in developed countries. In this context, the International Energy Agency (IEA) estimates that "... on average oil-importing developing countries use more than twice as much oil to produce a unit of economic output as do the OECD countries" (IEA, 2004). This high dependency on energy inputs, coupled with low efficiency in energy use, takes a toll on developing economies. Oil price shocks in net oil-importing countries are transmitted through a number of main channels:

First, the supply-side effect creates immediate economic distortions that hit oil-intensive production sectors. The supply-side effect refers to the reduced availability of a key production input (oil) when oil prices rise. Because the cost of other production inputs, notably labor, do not fall, the overall per unit cost of production rises, leading to reduced output levels. Since output prices do not necessarily rise with increasing oil prices, the profit margins of oil-intensive production sectors plummet and may have an overall negative effect on the macro economy.

Second, it is a well-established fact that a rise in oil prices leads to deterioration in the terms of trade of net oil-importing countries, and, subsequently, to a fall in the purchasing power of firms and households in net oil-importing countries (Dohner, 1981). This is essentially a transfer of wealth from net oil-importing to net oil-exporting countries. However, some argue that the effect of high oil prices can also be indirect, which works through the economies' trading partners. Increased trade between net oil-importers and net oil-exporters, where oil windfall is used to import more manufactured products from net oil-importing countries, may have a positive effect on the economies of net oil-importers (Abeysinghe, 2001). Therefore, the net effect of oil shocks on net oil-importing economies depends on how net-exporting countries decide to spend extra windfall purchasing power, and their trade preferences. Since most net oil-importing African economies are not well diversified or industrialized, their effective supply response capacities are limited, even if net oil-exporters choose to spend their windfalls on importing goods and services from them.

Third, increasing oil prices may lead to increased money demand in net oil-importing countries (Mork, 1994), and failure to meet this demand through increased money supply leads to higher interest rates. This has negative effects on consumption and investment, leading to lower growth. Consumption is affected through its positive relation to disposable income, and investment through increasing firm costs.

Finally, if oil price increases prevail over a long period, they may lead to a change in the production structure in favor of non-oil intensive sectors, which may lead to other distortions. The resulting reallocation of labor and capital across sectors in response to oil price increases can affect the unemployment situation in the long term (Loungani, 1986).

Overall, therefore, net oil-importing African countries remain vulnerable to energy price shocks, particularly non-export-oriented economies. Since economic diversification is still low in most African economies, energy shocks have the potential to continue taking a toll on the continent's economies. Given that the continent's energy use efficiency is among the lowest in the world—precisely at a time when energy prices are skyrocketing—and given the unique opportunity offered by discoveries of oil and gas fields on the continent, an explorative study of the oil and gas situation in Africa is timely, especially in the face of emerging evidence of the impact of the high level and volatility of oil prices.

1.2 Recent Developments

The price of crude oil rose from less than US$40.00 per barrel in 2004 to a high of US$70.85 per barrel in August 2005. Crude oil prices continued to rise, trading at about US$80.00 per barrel in 2006, and reaching a record of just over US$147.00 per barrel in July 2008. However, following the global financial crisis and the collapse of a number of banks and loss of jobs, demand for oil fell, leading to a drastic fall in the price of oil. By early November 2008, oil prices had plummeted to less than US$62.00 per barrel, settling at about US$42.00 per barrel by the end of February 2009. The causal factors of the earlier observed increase in oil and gas prices are many and vary widely from a market perspective. The main drivers of the high oil price can be summarized around basic demand and supply relationships.

On the demand side, the principal factor is the large and growing energy demand from especially China, India, and other emerging economies—driven by their rapid growth. Because most of the increased demand in global energy markets is a derived demand of these countries, and given the huge scope that exists for these economies to continue growing, there is a cautious expectation that energy prices will rise again in the medium term.

On the supply side, there has been a recorded sharp decline in the spare oil production capacity of OPEC, against a backdrop of noticeable decline in upstream investment by multinational oil companies, coupled with the prevailing crisis in the Middle East, a region that enjoys the world's largest oil and gas production capacity. With the current OPEC stance of not raising production levels, it is only reasonable to expect that the energy crisis will persist and that oil prices will rise again in the future.

In the short term, therefore, high energy prices will continue to hit net oil importers, which comprise about 71 percent of African countries. In particular, high oil prices are making it harder for African oil-exporting countries to meet critical development challenges, chiefly the Millennium Development Goals (MDGs). For instance, while such approaches as the Highly Indebted Poor Countries (HIPC) Initiative and the Multilateral Debt Relief Initiative (MDRI) have eased the debt burden on low-income African countries, the gains have been undermined by the oil crisis. In particular, net oil-importing HIPCs face difficulties as debt relief resources are diverted towards cushioning oil shocks.

A recent study by the African Development Bank (AfDB, 2008) concluded that "... the increase in the annual cost of oil between 2002 and 2006 is more than twice the projected savings from the debt relief initiatives" for low-income African countries. Thus, much of the gains from debt relief over the last decade are quickly being eroded or diverted to adjusting to oil price increases. This in turn has slowed down progress in delivering service in the social sectors as required by the debt relief initiatives, ultimately delaying progress in meeting the MDGs.

1.3 Africa's Oil and Gas Resources

Africa is well endowed with minerals, including fossil fuels and gas resources. However, knowledge about the quantities of these resources is limited and a comprehensive, country-based assessment still remains a challenge. New discoveries of oil and gas resources on the continent continue to emerge and present unique economic opportunities. However, the exploration and exploitation of these resources are yet to benefit the populations. Nigeria, for instance, has been exploiting oil resources for the last 50 years and is now the world's fourth largest oil exporter. Yet, its human and physical capital development is assessed to be 400 percent lower than it would have been if the oil revenues had flown into public funds, and if such funds had been utilized in the public interest to generate economic opportunities for all (African Development Report, 2007: 108–11). Issues of concern include technical limitations, inefficient contract negotiations, inadequate auctioning of extraction rights, inefficient taxation, and, most importantly, poor public expenditure prioritization and (lack of) transparency in the use of revenues.

The continent's poor management of its natural resources has been a recurring theme in recent debates and need not be repeated in detail here. While there is ample evidence of countries that have managed their natural resources (gold and diamonds, for example) in ways that have benefited their populations (Botswana, Namibia, South Africa), the continent has yet to see success stories in the case of oil and gas. However, with the ever-increasing discoveries of new oil and gas (for example, in Ghana, Tanzania, Mozambique and Uganda) and prospected fields in many countries (Sierra Leone, Mali, Kenya), the oil and gas sector still offers a unique opportunity that can be harnessed. It is critical for the continent to learn from the failures of the past in designing strategies that fully maximize the benefits of the new oil and gas discoveries.

It is equally important to explore ways to enhance regional energy security. In this context, an assessment of the oil and gas situation on the continent—the objective of this study—is a significant step towards harnessing the gains from oil and gas resources and achieving regional energy security.

1.4 Objectives of the Report

It is against this background that the African Development Bank (AfDB) and the African Union (AU) undertook joint studies on the *Impact of High Oil Prices on African Economies*—concluded in 2007—and an *Analysis of the Oil and Gas Situation in Africa,* whose findings are presented in this report. The findings of both studies will serve as technical inputs into the establishment of the proposed African Petroleum Fund (APF),[1] which is jointly developed by the AfDB and the AU. The idea of the APF first emerged from the Sixth General Assembly of the African Union (AU), held in Khartoum on January 24 2006, where the AU Commission was directed to convene a meeting of experts to consider a proposal for the establishment of an African Petroleum Fund (APF) within the African Union, to mitigate the effects of oil price increases on African countries.

The broad objective of the report is to shed light on key issues related to the social, environmental, and economic impacts of high and volatile oil and gas prices. This will be achieved by analyzing the oil and gas situation in Africa and examining the continent's major challenges and opportunities in the energy sector. The report seeks to achieve the following specific objectives:

- Undertake a comprehensive review and analysis of the oil and gas situation in Africa;
- Analyze Africa's most recent oil and gas performance and highlight the specific challenges and constraints

[1] The purpose of the envisaged African Petroleum Fund (APF) is to assist low-income, net oil-importing African countries severely affected by oil price increases. The goal is therefore to mitigate the balance of payments effects from the increase in oil prices on poor African countries dependent on oil imports. The specific objective of the APF is to smooth out domestic (African) price fluctuations in oil and petroleum derivative fuels caused by unpredictable changes in international market dynamics. The immediate expected result is reduction in the effects of unforeseen natural and geopolitical factors that drive international oil prices. In a wider perspective, the APF also aims to become a platform for the regional integration of African countries by promoting intra-regional import and export of oil products and by becoming a discussion forum for mitigating the impacts of high oil prices.

facing African countries in the exploitation and utilization of these resources;

- Discuss the impacts of rising oil and gas prices in African countries with *limited or no* resources (net importing countries);
- Analyze the unique role that development partners, including the African Development Bank and the African Union, can play through coordinated action to help African countries overcome challenges in the energy sector; and
- Make recommendations on future strategic directions to achieve optimal exploitation and utilization of the continent's oil and gas resources in a generationally equitable manner.

1.5 Scope

The report includes a comprehensive analysis of the oil and gas sector in Africa—background information that is pertinent to the implementation of the African Petroleum Fund (APF). It draws largely on the two APF-related studies conducted by the AfDB and the AU. The findings of these studies have been enriched through extensive literature review, desk research, and analysis of data from secondary sources.

Comprehensive analysis of the oil and gas situation in Africa, specifically, remains limited in the literature. The few existing regional studies—conducted by international financial institutions such as the World Bank and the IMF—categorize North Africa as a separate entity from Sub-Saharan Africa. This report uniquely highlights—through country examples and an African focus and perspective—the specific challenges and constraints facing the continent in the exploitation and utilization of its oil and gas resources. The report partly draws on a model that simulates the impact of high oil prices on African economies—a model developed by the Research Department of the Bank in a separate study. The roles of the AfDB and the AU are analyzed, considering their differing, but complementary, mandates geared towards the development of the continent. Finally, the report includes recommendations on future directions and actions for maximizing the benefits from Africa's oil and gas resources.

1.6 Outline

Chapter 1 provides the general background, objectives and methodology of the report. **Chapter 2** presents an overview of the status of oil and gas in Africa, including reserves, consumption and trade. **Chapter 3** analyzes how the continent has used its rich oil and gas resources historically, and how it can maximize the benefits from these resources going forward. **Chapter 4** discusses the impact of high oil prices on African economies, anchoring the analysis on a model developed by the AfDB. It also explores relevant recommendations and coping mechanisms to deal with high and volatile oil and gas prices from both the perspective of net oil-exporting and net oil-importing countries. **Chapter 5** discusses the main challenges stemming from the "resource curse" *and* opportunities for translating and converting the natural resource wealth into tangible benefits and economic growth. It further examines and highlights governance as the basic element of an enabling environment for transforming natural resources into shared growth. **Chapter 6** wraps up the report with recommendations; it discusses the main lessons learned and the role of development partners, including the AfDB and the AU, in achieving sustainable growth and energy security in Africa.

2

Oil and Gas in Africa

Africa has historically been referred to as a continent with rich natural resources but with high levels of poverty. Indeed, Africa is considered to be well-endowed with minerals, including fossil fuels, although the exact economic potential of these resources is not known. New discoveries of oil and gas in Africa—in locations never before thought to hold such resources in significant amounts—prove that the continent is still "virgin" in many aspects regarding exploration and exploitation of oil and gas resources.

2.1 The International Perspective

This chapter briefly reviews major developments in the oil and gas sector in the context of developments in global energy markets over the last 40 years. The discussion focuses on the dynamics of oil and gas markets over the last ten years, during which Africa has been playing an increasingly important role.

2.1.1 *Evolution of World Energy Markets*

The evolution of world energy markets in the post-1970 period has been dramatic and its impact on the world economy and on politics profound. This is illustrated by the worldwide economic ripple effects caused by the volatility and occasional spectacular spikes in the prices of dominant global energy resources such as oil and gas. Another major trend with significant impact has been the fundamental changes in the structure, conduct, and performance of the oil and gas sector—including considerable improvements in oil and gas technology, unprecedented consolidation among multinational oil companies, increasing global price transparency implicit in oil trade, new market fundamentals, and environmental considerations. The transition from a seller's market environment (in the early 1970s), first controlled and managed by multinational oil companies and subsequently by major net oil-exporting countries (OPEC), to a buyer's market illustrates the profound changes that occurred in the oil industry during the period spanning the 1970s to the late 1990s. These issues are further explored in the following section, which focuses on oil prices and on post-1990 events in particular.

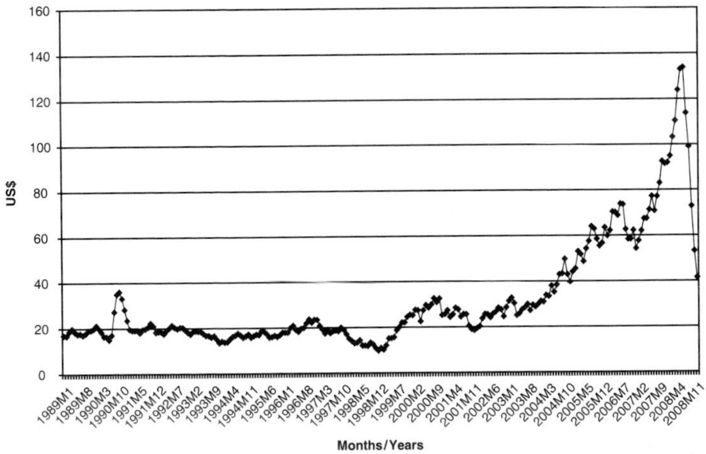

Figure 2.1: Crude Oil Spot Price between 1989 and 2008

Source: Authors, from IMF Commodity Prices Database, January 2009.

2.1.2 *Trends in Prices of Crude Oil and Petroleum Products*

Crude oil prices exhibit wide swings in times of shortage or excess supply. The crude oil price cycle (Figure 2.1) may extend over several years, even decades, responding to changes in overall geopolitical developments and to demand and supply trends. Understanding and distinguishing among the factors affecting short- and long-term behavior has been the subject of much discussion since the 1970s. The stylized facts about global energy have been widely documented; however, the scope of this publication does not allow further elaboration on the debate. This report briefly examines the recent dynamics in the industry, as highlighted by price changes and industry structure, conduct, and performance. It also draws some emerging lessons from these developments.

The dramatic rise in crude oil prices in 1973–74 marked a new era in the industry. The rise in oil prices was triggered in part by the supply shortage associated with the oil embargo imposed on some major oil-consuming countries—as a result of the Middle East crisis—and the transfer of oil property rights from multinational oil companies to oil-producing countries. It is worth noting that oil price increases in 1973–74 and 1979–80 signaled the end of the era of inexpensive energy in general. The nominal price of oil, which was below $2 per barrel in 1970, rose to over $40 in 1981, to just over $147 in July 2008 and to about $42 per barrel in February 2009. Indeed, since the beginning of this decade, and over the past two years in particular, world oil and gas prices have risen to new historic heights. Predictably, the prices of petroleum products have also skyrocketed, triggered partly by the development in crude oil prices.

The prices of natural gas imports, which averaged less than $2 per thousand cubic feet in the 1990s, jumped to $4 in 2001, and doubled to an annual average of $8 in 2005. Overall, large increases in gas prices in the past decade ushered in a new era with considerably higher energy prices. However, because natural gas does not have the same political profile as crude oil, its market development has been less contentious. Natural gas prices, like oil prices, have also experienced volatility since 1990. For example, the price of U.S. natural gas imports fell by about 37 percent in September 2008, from a peak of $11.99 in October 2005.

2.1.3 *Market Dynamics: Emerging Trends*

The behavior of crude oil prices can be viewed from two broad perspectives: prior to 1973 and post 1973. During the first period, prices exhibited long-term stability. The multinational oil companies in control of the industry at that time managed market supply of crude to avoid substantial excess supply that could destabilize the market. The playing field has changed significantly since the 1970s. A series of events were fundamental to the radical changes in market structure. First, oil property rights were transferred from multinational oil companies to OPEC oil-producing countries. Second, the oil price increases of the 1970s, coupled with extremely high taxes on oil in the major OECD consuming countries, helped trigger demand and supply responses that eventually produced the first major market slump in the 1980s. However, increasingly in recent years, the supply response to increased demand has been weak, inducing higher price volatility.

In recent years, several factors have led to the current state of the oil market, which is characterized by volatile and high prices for crude oil and petroleum products. These factors include:

- Rise in demand in emerging economies, especially in China and India;

- Decline in the spare capacity of major producing countries and the peaking of production in several important oil-producing areas (such as the North Sea and North America);
- Decline in global investment in the industry;
- Lack of expansion in refinery capacity;
- Supply bottlenecks and uncertainties associated with domestic problems (for example, in Nigeria), and international politics (Iran, Iraq, Venezuela, and Russia) that impact supply;
- Supply uncertainties associated with extreme weather events (such as hurricanes);
- Lack of a dominant actor in the market to manage excess supply and demand; and
- Commoditization of world oil (for example, through NYMEX).

In the longer term, there will primarily be three market drivers: demand from emerging countries (China and India in particular); production from OPEC countries; and inventory movement in major consuming countries (especially the United States). Increasingly, environmental policy considerations may also shape the future of the market. Policies aimed at reducing carbon emissions in the quest to mitigate global warming are likely to impact the future use of oil.

Meanwhile, price movements—reflecting market conditions—as well as pure speculation will continue to have a significant impact on the global economy.

Sustained energy price shocks, as witnessed in recent years, constitute a major constraint on economic growth. Since oil and gas are likely to remain the dominant energy fuels, recent developments in these markets have far-reaching implications for achieving universal, affordable, commercial energy access in net oil-importing developing countries. Undoubtedly, for the more than two billion poor people living in Africa, Asia and Latin America, the current oil and gas market dynamics are likely to make universal and affordable access to commercial energy more difficult to achieve. The impact of rising oil prices for Africa is further analyzed in Chapter 4.

2.1.4 *Supply and Demand Outlook*

Any analysis of the outlook of oil and gas supply and demand is fraught with difficulties. Assumptions have to be made about growth in demand, commodity prices, reserve base, potential new discoveries, technology evolution, and world politics. The problem is that data on production and demand from the recent past, as well as predictions in the short to medium term, are often inaccurate. These inaccuracies stem from the way the data is compiled using OPEC production estimates as well as commercial figures from other countries around the world. These figures are sometimes slanted for political or economic reasons. The uncertainty increases for predictions of future production demand. In the near term, companies are constantly revising their production targets downwards, and in the longer term, published supply and demand estimates show a large margin of error. Recent analysis by ExxonMobil, for instance, indicates that, over the next 10 years, oil and gas demand will increase by about 2 percent a year, while current fields in production will deplete at an average of 3 percent to 5 percent per year. A similar situation exists for gas production, demand, and supply, although the available reserves figures are even less certain than those for oil. A significant amount of gas is still in the "possible" category.

2.1.5 *Global Energy Resource Profile: Oil and Gas in Perspective*

World proven crude oil reserves[1] at the end of 2006 were estimated at 1,208.2 billion barrels.[2] Global oil reserves are unequally distributed as the data on regional

[1] Conventional crude oil reserves include all crude oil that it is technically possible to produce from reservoirs through a well bore, using primary, secondary, improved, enhanced, or tertiary methods. This does not include liquids extracted from mined solids or gases. Oil reserves are classified as proven, probable, and possible. Proven reserves are generally intended to have at least 90% or 95% certainty of containing the amount specified. Probable reserves have an intended probability of 50%, and possible reserves an intended probability of 5% or 10%. Current technology is capable of extracting about 40% of the oil from most wells. In several major producing countries, the majority of reserves claims have not been subject to outside audit or examination. Unconventional sources, such as heavy crude oil, tar sands, and oil shale, are not counted as part of oil reserves. For further elaboration on reserve categories and terminology, also see Whateley and Harvey (1994).

[2] The analysis in this chapter is based mainly on data from *BP Statistical Review of World Energy* (2006 and 2007 volumes).

composition of proven reserves in Table 2.1 demonstrates. The Middle East region is the dominant region, accounting for 61.5 percent of estimated world proven oil reserves. Europe and Eurasia come in a distant second, with 12.0 percent; while Africa ranks a close third, with 9.7 percent of global reserves. The geographical distribution has remained relatively unchanged in the past three decades. The OECD region, which includes the major oil- and gas-consuming countries in the world, accounts for less than 6.6 percent of global reserves. In contrast, OPEC holds 75.7 percent of global reserves.

From a longer-term perspective, world oil reserves grew at an annual rate of 1.7 percent between 1980 and 2006. Reserves growth in Africa was double the global rate, at 3.2 percent; while OPEC reserves grew 1.5 percent.

The upward trend in proven reserves is attributable to two main factors: rising oil prices and technological progress. It is widely recognized that the upward trend in demand and crude oil prices have provided a very attractive environment for exploration and for a subsequent increase in proven and probable oil and gas reserves. Technological progress has boosted average recovery rates and also enabled economically attractive production from what was previously regarded as marginal and infra-marginal oil fields.

The lopsided distribution of oil wealth has generated concerns about global access, oil geopolitics, and energy security. In fact, the concentration of proved reserves in a few countries has been a contentious factor in the evolving state of global energy markets in past decades.

Table 2.1: Proved Crude Oil Reserves by Region, 1980 to 2006 (Billion Barrels)

Region	**1980**	**1985**	**1990**	**2000**	**2005**	**2006**	**% of total in 2006**
Africa	53.3	57.0	58.7	93.4	114.3	117.2	9.7
North America	92.5	101.5	96.3	75.6	60.7	59.9	5.0
South and Central America	26.7	62.9	71.6	97.9	103.2	103.5	8.6
Europe & Eurasia	98.4	78.6	80.3	114.1	145.2	144.4	12.0
Middle East	362.4	431.3	657.7	691.0	742.7	742.7	61.5
Asia Pacific	33.8	39.1	36.3	42.6	40.5	40.5	3.4
World	667.1	770.4	1,000.9	1,114.7	1,209.5	1,208.2	100
of which OECD	109.2	118.6	115.1	100.0	**81.9**	79.8	6.6
OPEC	434.6	535.8	765.9	840.5	**914.5**	914.6	75.7
Non-OPEC	150.5	172.0	171.7	180.8	**176.4**	174.5	14.4

Data Source: BP Statistical Review of World Energy 2007 (BP, 2007).

World proven gas reserves in January 2007 were estimated to be 181.46 trillion cubic meters. However, although the distribution of global gas resources is unequal, similar to that of crude oil, its regional imbalance is less pronounced. Of the world's proved reserves, the Middle East region ranks first with a 40.5 percent share (Table 2.2); Europe, including gas reserves in Russia and the former Soviet Republics, follows closely with 35.3 percent; while Africa ranks fourth, with 7.8 percent.

2.1.6 *Trends and Patterns of Global Energy Production and Consumption*

To understand recent energy market developments, it is necessary to examine the long-term trends and patterns of world primary energy supply and demand. The analysis is based only on commercial and conventional energy sources due to lack of data.

Table 2.3 shows that world primary energy supply rose from 288 quadrillion BTU (British Thermal Units) in 1980 to 443 quadrillion BTU in 2004. Primary energy supply almost doubled in Africa. Table 2.4 presents the geographical distribution of primary energy production. It is evident that oil and gas remain the dominant source of energy worldwide.

The regional composition of global energy consumption reveals a wide disparity in global access to reliable and adequate commercial energy. Although Africa (Sub-Saharan Africa and North Africa) accounts for about 15 percent of the world's population, it consumes only

Table 2.2: Regional Distribution of Natural Gas Reserves, 1980 to 2006 (Trillion m^3)

Region	1980	1990	1995	2000	2005	2006	% share of total in 2006
Africa	5.99	8.55	9.93	12.47	14.08	14.18	7.8
North America	9.89	9.49	8.47	7.49	7.83	7.98	4.4
South and Central America	2.78	5.25	5.96	6.98	6.85	6.88	3.8
Europe & Eurasia	36.00	59.84	63.16	61.74	64.30	64.13	35.3
Middle East	24.69	37.99	45.37	59.81	72.49	73.47	40.5
Asia Pacific	4.47	9.88	10.54	12.28	14.66	14.82	8.2
World	83.30	131.00	143.42	160.76	180.20	181.46	100.0

Data Source: *BP Statistical Review of World Energy 2007* (BP, 2007).

Table 2.3: World Primary Energy Production in Quadrillion BTU, 1980–2004

Year	Africa	North America	S. & C. America	Europe	Eurasia	Middle East	Asia & Oceania	World Total
1980	**17.395**	83.276	12.083	40.216	56.463	42.265	35.861	287.559
1990	**21.610**	91.903	16.749	46.907	72.111	40.995	59.389	349.663
2000	**27.844**	98.754	26.042	50.821	55.689	57.480	80.499	397.129
2001	**28.098**	99.468	25.990	51.412	57.705	56.155	85.467	404.295
2002	**28.004**	100.019	25.318	51.200	59.463	54.251	87.867	406.121
2003	**30.105**	98.523	25.738	50.672	63.209	57.612	95.855	421.714
2004	**32.043**	99.310	27.169	50.610	66.714	62.078	105.177	443.100

Data Source: U.S. Department of Energy (2006*a*).

Table 2.4: World Primary Energy Production by Fuel Type, 1980–2004 (% of Total)

	Coal	Natural Gas	Natural Gas Liquids	Crude Oil	Nuclear Power	Hydro Power	Geothermal and Others	Total
1970	29.2	17.2	1.7	45.1	0.4	5.6	0.7	100.0
1980	24.8	19.0	1.8	44.5	2.6	6.2	1.0	100.0
1990	26.0	21.7	2.0	37.0	5.8	6.4	1.1	100.0
2000	23.0	23.0	2.5	36.9	6.5	6.8	1.3	100.0
2004	25.6	23.1	2.6	34.9	6.2	6.2	1.4	100.0

Data Source: U.S. Department of Energy (2006*a*).

3 percent of global commercial energy. One can reasonably assume that the region's low consumption and share of energy reflect its low access to affordable commercial energy as well as its low level of development and industrialization. In contrast, Africa's share in global energy production is about 12 percent and trending upwards (this paradox is discussed extensively in subsequent sections of this report).

The global dependence on oil and gas is evident in Table 2.5. Of the 10.5 billion metric tons of oil equivalent of world energy consumption in 2005, oil remains the most important source of energy, accounting for 36 percent of total consumption. Coal comes second, with a 27.8 percent share; and natural gas third, with a 24 percent share. The combined share of oil and gas represents 60 percent of global energy consumption, illustrating the dominance of hydrocarbon fuels in world energy demand. Remarkably, the three non-renewable

Table 2.5: Global Dependence on Oil and Gas, 2005 (%)

	Oil	Natural Gas	Coal	Nuclear Energy	Hydro Electric	Total
Africa	40.9	20.3	31.7	0.9	6.3	100.0
North America	40.4	24.9	21.9	7.5	5.3	100.0
South and Central America	44.5	22.3	4.2	0.7	28.3	100.0
Europe and Eurasia	32.3	33.8	18.0	9.6	6.3	100.0
Middle East	53.2	44.3	1.8	–	0.8	100.0
Africa	40.9	20.3	31.7	0.9	6.3	100.0
Asia	32.6	10.7	48.1	3.7	4.9	100.0
World	36.4	23.5	27.8	6.0	6.3	100.0
United States	40.4	24.4	24.6	8.0	2.6	100.0
EU 25	40.8	24.7	17.4	12.9	4.1	100.0
OECD	41.0	23.0	21.1	9.6	5.4	100.0

Data Source: *BP Statistical Review of World Energy* (BP, 2006).

fossil fuels—oil, natural gas, and coal—constitute almost 90 percent of commercial energy consumed worldwide.

2.2 Oil and Gas and the Global Economy

The interrelationship between oil and gas and the economic status and development of countries remains a dominant and persistent theme. A key factor is the fact that the world economy runs largely on hydrocarbons. The issue of unequal global access to commercial energy and the gains of global economic prosperity have also frequently been highlighted in international discourse.

Energy has traditionally played an important role in the economy of any nation because it is an indispensable input for economic growth and development. Two-thirds of global energy requirements are met from oil and gas supplies. Conventional wisdom holds that there is a strong correlation between energy consumption per capita and the level of economic and social progress. This still holds true despite globalization, increasing growth—in Asia, for example—and energy market developments in recent decades.

Since economic and social development depend, in large part, on an adequate supply of energy at affordable prices, changes in energy prices are bound to affect economic growth and development. However, the resilience of developed and developing countries to the shocks produced by these changes varies as a function of the different economic strengths of these countries. Affordable access to adequate and reliable energy supply was a central factor in the achievement of the large productivity gains that guaranteed sustained economic growth and led to a significant leap in material prosperity and widespread improvement in human well-being in the last century. The data in Table 2.6 supports the hypothesis that there is a strong correlation between energy (and oil) consumption per capita and the level of economic and social progress. The data clearly shows that countries with higher per capita income also consume more energy per head.

Table 2.7 further reveals the wide disparities in global access to reliable and adequate commercial energy as well as in the sharing of the fruits of global economic prosperity. Sub-Saharan Africa (SSA), with almost 12 percent

Table 2.6: Per Capita Energy Consumption and Income for Selected Countries

Country	Per Capita Energy Consumption 2004 (BTU)	Per Capita Income 2005 (GNI)	Country	Per Capita Energy Consumption 2004 (BTU)	Per Capita Income 2005 (GNI)
Norway	424.0	59,590	Egypt	33.1	1,250
United States	342.7	43,740	Cameroon	5.2	1,010
Japan	177.7	38,980	Congo Rep	4.9	950
United Kingdom	166.5	37,600	India	14.5	720
France	186.1	34,810	Senegal	5.8	710
Germany	178.3	34,580	Sudan	3.8	640
Hong Kong	159.1	27,670	Nigeria	8.1	560
Singapore	444.6	27,490	Kenya	5.3	530
Kuwait	470.0	24,040	Zambia	11.1	490
United Arab Emirates	925.4	23,770	Ghana	6.6	450
Korea, South	185.5	15,830	Burkina Faso	1.4	400
Saudi Arabia	236.5	11,700	Chad	0.3	400
Seychelles	147.7	8,290	Mali	1.0	380
Libya	133.0	5,530	Guinea	2.4	370
Mauritius	45.0	5,260	Central African Republic	1.4	350
Botswana	32.6	5,180	Tanzania	2.0	340
Gabon	29.4	5,010	Mozambique	7.3	310
South Africa	115.2	4,960	Uganda	1.5	280
Malaysia	107.1	4,960	Niger	1.4	240
Namibia	27.5	2,990	Rwanda	1.5	230
Tunisia	33.4	2,890	Guinea-Bissau	3.8	180
Algeria	38.6	2,730	Ethiopia	1.2	160
China	45.9	1,740	Liberia	2.6	130
Morocco	13.8	1,730	Burundi	1.0	100
Angola	12.2	1,350			

Data Sources: World Bank Data for Gross National Income (WB, 2006*e*); EIA (2006), and U.S. Department of Energy (2006*a*).

Table 2.7: Inequities in Global Income, Oil Consumption, and Population (% Share)

Region	Population	Gross National Income	Oil Consumption Per Capita
Sub-Saharan Africa (SSA)	11.5	1.2	3.4
Latin America	8.6	4.5	5.8
Middle East & North Africa	4.7	10.6	7.1
North America	6.7	32.8	29.5
High Income	15.7	78.9	59.2*

* OECD's Oil Consumption Share.
Data Sources: *World Development Report 2007*: *Income Data* (WB, 2007); and BP (2006).

of the world's population in 2005, accounted for only 1.2 percent of world economic output and for 3.4 percent of global oil consumption. In comparison, North America, with about 7 percent of the world's population, accounted for 33 percent of economic output and 30 percent of oil consumption.

2.2.1 *Impact of Oil and Gas Prices on the Global Economy*

Historically, there has been a strong relationship between oil and gas prices and global economic growth. Indeed, economic and other policy responses to changing global energy market conditions have had a far-reaching impact on the economies of both developed and developing countries.

From a *microeconomic perspective*, changes in the price of a commodity result in substitution and income effects. In a normal case, the real income effect reinforces the pure substitution effect and yields a negative relationship between the amount purchased and the price of the product. Higher prices lead to reduced consumption, all other things being equal.

Analysis of the *macroeconomic* effects of higher oil prices has been a recurrent theme since the first major oil price shock of the 1970s. Historical evidence shows that oil price increases and volatility have had a significant impact on global, regional, and national economic performance and outlook. Conventional wisdom, backed by empirical analysis, suggests that increases in oil prices tend to exert strong inflationary pressures, reduce output, slow down economic growth, and exacerbate unemployment in net oil-importing countries. Thus, the rising costs of oil imports caused by both the exchange rate depreciation and the rising dollar price of oil are bound to have profound ripple effects throughout national economies.

Broadly speaking, higher oil prices can impact aggregate economic activity directly or indirectly. The main channels through which oil prices impact macroeconomic aggregates are well understood (IMF, 2000, 2005). These channels include the following:

1. Terms of trade effects: These result in the transfer of income from oil-importing countries to oil-producing countries at the global level and from oil consumers to oil producers at the domestic level. At the global level, the net gain to oil-exporting countries is offset by the net loss to net oil-importing

countries. The bigger the oil price increase, the larger the potential macroeconomic effects are likely to be at the national, regional, and global levels.

2. Input price: This channel works through production and investment via increasing cost of production in the economy. The size of this impact of higher oil prices on the economy depends on oil intensity of aggregate output. Adding price volatility to the price shock, the consequences are high cost of production, deterrence of irreversible investment, and ultimate reduction in aggregate economic activity, or, at best, slowdown in rate of economic growth.
3. Exchange rate pass-through: The invoicing currency for the oil trade has traditionally been the U.S. dollar. Oil-importing countries are adversely affected by changes in the value of the U.S. dollar vis-à-vis other currencies. The exchange rate pass-through effect exacerbates the input shock effect of oil prices in net oil-importing developing countries. The magnitude of the pass-through depends on government policy and the state of energy market liberalization. The size of the pass-through also depends on the structure of the domestic oil market. A more competitive market structure would minimize the pass-through relative to monopolistic or oligopolistic cartel-like control of the market.
4. Oil price pass-through to inflation: Oil price increases work as a shock to the aggregate price level, exacerbating inflationary pressure that

decreases money balances with its macroeconomic consequences. There could be partial or total pass-through from higher oil prices to core inflation depending on the monetary policy stance.

5. Balance of payments impact through higher oil import costs: This will be more severe if there are no countervailing and compensating changes in export revenues through higher volume of exports and through terms of trade gains.

The magnitude of the macroeconomic impact of higher oil prices has been the subject of considerable debate in the literature (Mork, 1989; Hamilton, 1996; and Lee, Ni and Ratti, 1995). Two basic methodological approaches have been adopted in quantifying the macroeconomic impact of higher oil prices. The macro-econometric approach is the most widely used methodology. From this approach, substantial empirical evidence shows that both developed and developing countries have been able to cope increasingly better with the challenges associated with oil price increases with each successive decade since the first major price increase in the 1970s (IMF, 2000). Nevertheless, economic performance would have been more robust if oil prices had not risen so sharply.[3]

[3] Two major defects of the conventional macro-econometric approach are worth noting. The first is its inability to capture the non-linear and asymmetric relationship between oil price changes and output and employment. Oil price volatility, as evident in the time series behavior of oil prices, further complicates the problem of estimating the magnitude of the macroeconomic impact of higher oil prices. The second defect of the macro-econometric approach is its inability to capture properly the distributional effects of higher oil prices. Although this may not be an

Finally, an important issue is the role of economic policy—at both the micro and macro levels—in mitigating the potential adverse economic impact of higher oil prices. Two questions come to mind: What can and should be done to mitigate or even neutralize the impact of higher oil prices? In other words, what is the optimal policy mix that would help economies cope with the high economic and social costs of the external shocks that accompany oil and gas market price changes? It is in this context that an *African Petroleum Fund* was proposed to assist African countries to deal with oil price shocks.

2.2.2 *Importance of Petroleum Products*

The analysis in this section focuses on the magnitude and importance of petroleum products for society and for fueling the world economy. There are strong links between the oil industry and other sectors of the domestic economy.

2.2.2.1 PETROLEUM PRODUCTS SUPPLY CHAIN

The energy supply cycle can be divided into upstream and downstream elements: the upstream comprises exploration and production activities; while the downstream includes refining and retail activities. In basic terms, the cycle consists of acquisition of crude oil,

important issue in developed countries, the same cannot be said of low-income countries, where a majority of the people lives on less than $1 a day.

transportation to the refinery, refining, transportation to storage, management of inventories, and selling of final products. Specifically, the complex downstream industry consists of

- Refining: Crude oil ships and refinery terminals
  - Refinery storage and local terminal rack
  - Refinery products
- Wholesale: Shipping activities
  - Marine terminals
  - Local and international markets
- Retail: Bulk terminal, retail/wholesale, and industrial and home usage.

2.2.2.2 IMPORTANT REFINED PRODUCTS

Refineries turn crude oil into liquefied products, aliphatic chemicals, asphalt paving mixtures, aviation fuels, benzene, butylenes, resid/fuel oil/bunker, cumene, cyclic aromatic hydrocarbons, diesel fuels, ethylene, gasoline, heating oils, hydraulic fluids, jet fuels, kerosene, lubricating oils and greases, naphtha, naphthenic acids, paraffin waxes, petrochemicals, petroleum coke, petroleum jelly, petroleum lubricating oils, propane gases, propylene, solvents, styrene, tar, toluene, and xylene (EIA, 2008).

Petroleum products, especially motor gasoline, diesel, and jet fuel, provide virtually all the energy consumed in the transportation sector. Transportation is the greatest single use of petroleum, accounting for over 67 percent of all U.S. petroleum consumed in 2005. The industrial sector is the second largest petroleum-consuming sector and accounts for about 24 percent of all petroleum consumption in the U.S. The residential/commercial and electric utility sectors account for the remaining 9 percent of petroleum consumption (EIA, 2008).

Distillate fuel oil includes diesel oil, heating oils, and industrial oils. It is used to power diesel engines in buses, trucks, trains, automobiles, and other machinery. It is also used to heat residential and commercial buildings and to fire industrial and electric utility boilers.

Liquefied petroleum gases (LPGs) rank third in usage among petroleum products, behind motor gasoline and distillate fuel oil. LPGs are used as inputs (feedstock) for petrochemical production processes. This is their major non-fuel use. LPGs are also used as fuel for domestic heating and cooking, farming operations, and as an alternative to gasoline for use in internal combustion engines. Most jet fuel is a kerosene-based fuel used primarily in commercial airlines. Naphtha jet fuel meets the specifications required for certain military aircraft. Kerosene-type jet fuel is sometimes blended into heating oil and diesel fuel during periods of extreme cold weather. Although this sector uses relatively little petroleum, compared with the transportation and industrial sectors, the electric utility sector depends on petroleum for about 3 percent of its total energy requirements. Residual fuel oil is also used as bunker fuel (fuel for ships), industrial boiler fuel, and heating fuel in some commercial buildings. Kerosene is used for residential and commercial space heating. It is also used in water heaters, as a cooking fuel, and in lamps. Petroleum coke can be used as a relatively low-ash solid fuel for power plants and industrial use (marketable coke) if its sulfur content is low enough, or used in non-fuel applications (catalyst coke), such as in refinery operations (Nageria, 2005).

Non-fuel use of petroleum is small, compared with fuel use, but it is nevertheless very important from a commercial and economic point of view. Non-fuel uses for petroleum include various specialized products for use in the textile, metallurgical, electrical, and other industries. A partial list of non-fuel uses for petroleum includes (Nageria, 2005):

- Solvents such as those used in paints, lacquers, and printing inks;
- Lubricating oils and greases for automobile engines and other machinery;
- Petroleum (or paraffin) wax used in candy, packaging, candles, matches, and polishes;
- Petrolatum (petroleum jelly) in medical products and toiletries;
- Asphalt used to pave roads and airfields, to surface canals and reservoirs, and to make roofing materials and floor coverings;
- Petroleum coke used as a raw material for many carbon and graphite products, including furnace electrodes and liners, and the anodes used in the production of aluminum; and

- Petroleum feedstock used as chemical feedstock derived from petroleum, mainly for the manufacture of chemicals, synthetic rubber, and a variety of plastics.

Petroleum has been used as feedstock in the production of petrochemicals since the 1920s. Petrochemical feedstocks also include products recovered from natural gas and refinery gases (ethane, propane, and butane). Petrochemical feedstocks are converted to basic chemical building blocks and intermediates, such as ethylene, propylene, normal and iso-butylenes, butadiene, and aromatics such as benzene, toluene, and xylene, which are in turn used to produce plastics, synthetic rubber, synthetic fibers, drugs, and detergents.

2.2.2.3 INDIRECT IMPORTANCE OF PETROLEUM PRODUCTS

Petroleum products are very important for governments, essentially because they are a source of revenue through tax contributions. Furthermore, both downstream and upstream industries and associated services employ a significant number of people. The oil industry has social responsibilities with local communities where it helps in building access, schools, and hospitals. These aspects are further analyzed in Chapters 3 and 4.

2.2.3 *Natural Gas and Biofuels*

The use of petroleum products has been the subject of public outcry because of their environmental impact.

There is also growing concern about the viability of oil-based fuels and products—leading to initiatives to find and develop alternative energy sources.

At present, alternative energy sources—to the oil-based fuels and products—are quite limited and have until recently consisted of gasoline blends, ethanol, methanol, and liquefied gas. Most of these products still rely on expensive technology, which makes them less competitive when compared to oil products. The alternative energies are still held back by the extremely large capital installation costs, which reduces its popularity in today's world economy. However, both natural gas and biofuels in various forms have gained considerable attention and momentum in recent years.

2.2.3.1 NATURAL GAS

Natural gas can be used in a number of ways: in households, in industry, and for power generation. It can also be used in the transport sector, in the form of compressed natural gas (CNG), as a substitute for gasoline and diesel. There has been a steady shift towards natural gas as a clean (that is, cleaner than oil and coal) and relatively cheap source of power in recent years. However, by virtue of being a gas, natural gas is more costly to transport and store than liquid or solid fuels. Consequently, natural gas transportation is more complex and costly because it requires special vehicles or special pipelines. To address this problem, economies of scale must be pursued, especially in laying down extensive pipelines and distribution systems. A pipeline network is economically viable especially if there are large consumers (such as power plants and large industrial operations). The factors needed for sustainable conversion to CNG include the existence of a gas distribution pipeline for other natural gas users, close proximity to the supply, and inter-fuel taxation policy.

Biofuels are derived from biological (for example, agricultural) sources (Chow, 2007):

i. Cereals, grains, sugar crops and other starches can be fermented to produce ethanol, which can be used pure or as a blending component (as ethanol or after being converted to Ethyl Tertiary Butyl Ether, ETBE;[4]

ii. Cellulosic materials, including grasses, trees and various waste products from crops, wood processing facilities, and municipal solid waste, can also be converted to alcohols or diesel type fuels;

iii. Oil-seed crops (such as rapeseed, soybean, and sunflower) can be converted into methyl esters, which can be blended with conventional diesel or burnt as pure biodiesel;

[4] ETBE is an oxygenated fuel that can be blended with gasoline to make it burn more cleanly and thus improve overall air quality. ETBE is produced by mixing ethanol and isobutylene and reacting them with heat over a catalyst. The promise of ETBE is that it eliminates many of the historical impediments to the greater use of ethanol, such as increased volatility of gasoline and incompatibility with gasoline pipelines. This would allow ETBE to be used at the refinery level and be economically transported to areas that previously had not been able to utilize ethanol (Governors' Ethanol Coalition, 2007).

iv. Organic waste material can be converted into energy forms, which can be used as automotive fuel: waste oil (such as cooking oil) into biodiesel; animal manure and organic household wastes into biogas (such as methane); and agricultural and forestry waste products into ethanol.

Countries such as Argentina, Brazil, Canada, Colombia, India, Malaysia, Mozambique, the Philippines, Thailand, and the United States of America have been promoting and boosting the production and use of biofuels (biodiesel from soy bean palm and jatropha; ethanol from sugar cane and others) and LNG as a coping mechanism for high oil prices. Most of these countries have adopted targets for increasing the contribution of biofuels to their transport fuel supplies. In terms of production costs, ethanol from sugar cane has historically been the lowest-cost biofuel. However, recent surges in world sugar prices have altered the relative economics of biofuels, with biodiesel from palm oil emerging as potentially more viable under certain market conditions (Chow, 2007).

2.3 Africa's Oil Reserves and Production

2.3.1 *Reserves*

Reserves are defined as the known and estimated quantity in units of oil or gas that can be produced. The certainty to which we know the quantities defines the category of reserves. As outlined in Section 2.1, there are proven, probable, and possible reserves. It should thus be noted that a reserve estimate is dynamic and can change depending on the current technology level, economic conditions (for instance, the price of selling oil versus the costs of extracting it), location, and associated environmental issues.

The most common notion of "proved reserves"—declared quantities known to a high level of confidence in terms of primarily geological data—is used in this report. However, it is worth noting that even with this presumably physical and measurable estimation, data on oil and gas resources from geological resources may be inaccurate for a number of reasons, including: weak government capacity to monitor oil companies (which may not wish to have an accurate data disclosure); geopolitical interests of governments (who may not want a full disclosure either); or lack of geological survey data.

Figure 2.2 presents an overview of proved oil reserves in Africa. The majority of African oil reserves (and production) is located in Libya, Nigeria, Algeria, Angola, and Sudan, which together account for more than 90 percent of the continent's reserves. There is growing knowledge about Africa's proved oil reserves, which rose from 53.3 billion barrels in 1980 to 117.2 billion barrels in 2006, a share of 9.7 percent of total world reserves. The largest reserves are in Libya and Nigeria, which account for 3.4 and 3.0 percent, respectively, of world reserves (BP, 2007).

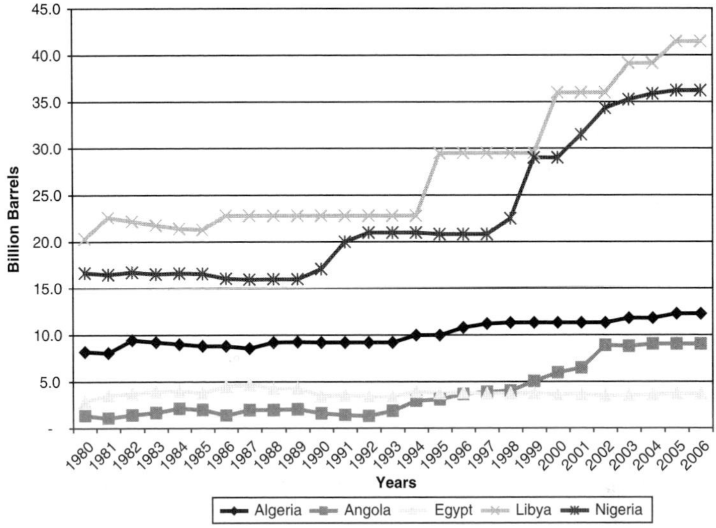

Figure 2.2: Proved Oil Reserves in Africa (Billion Barrels)

Data Source: Authors, with data from *BP Statistical Review of World Energy* (BP, 2007).

Table 2.8 presents an overview of the most important oilfields and Map 2.1 shows all known major deposits of oil, gas, and coal resources in Africa.

2.3.2 *Production*

Africa is an important player in world oil production, with a total share of 12.1 percent in 2006. The major producers are Nigeria, followed by Algeria, Libya, Angola, Egypt, Equatorial Guinea, and Sudan in succession (Figure 2.3). The four major producers in Africa (Nigeria, Algeria, Libya, and Angola) together account for

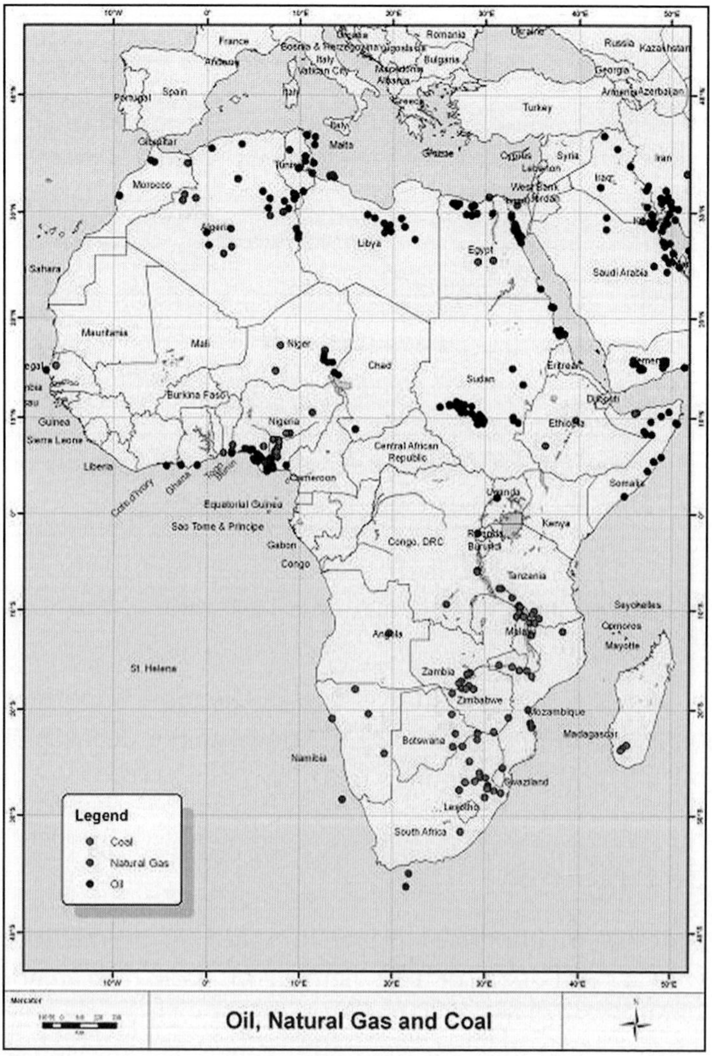

Map 2.1: Oil, Gas, and Coal Reserves in Africa

Source: Council for Geoscience (2007).

Table 2.8: Major African Oil Fields, 2007

State	Deposit Name	Size of Resource (Barrels)	Status
Algeria	Hassi Messaoud North & South	>300 million	Continuously producing
Algeria	Rhourde El Baguel; Hassi Berkine South; ZarzanTine; and Edj'leh	All 4 resources: 160–300 million	Continuously producing (all)
Angola	29 offshore and onshore blocks	Several resources >160 million within blocks	Some producing
Chad	Doba	160–300 million	Deposit never exploited
Egypt	El Morgan; and July Oilfield	160–300 million	Continuously producing
Gabon	Emeraude; Loango	>300 million	
Libya	Zelten	160–300 million	Derelict mine
Libya	Waha; Amal; Serir; Gialo; and Dahra	All resources: 160–300 million	Continuously producing (all)
Morocco	Meskalia	160–300 million	Continuously producing
Nigeria	Usan; Ukot; Aparo; Agabami; and Bonga Sw	All resources: 160 million	Continuously producing (all)
Nigeria	Jones Creek	160–300 million	Continuously producing
Tunisia	El Borma	160–300 million	Continuously producing

Data Source: Council for Geoscience and Mintek (2007).

77 percent of the continent's production and contribute 9.2 percent to world oil production.

Figure 2.3 illustrates the trend of oil production in Africa. It is worth noting that Libyan production dropped sharply from 3,300 barrels per day (bpd) in 1970 to 1,514 bpd in 1975. At the same time, Nigerian

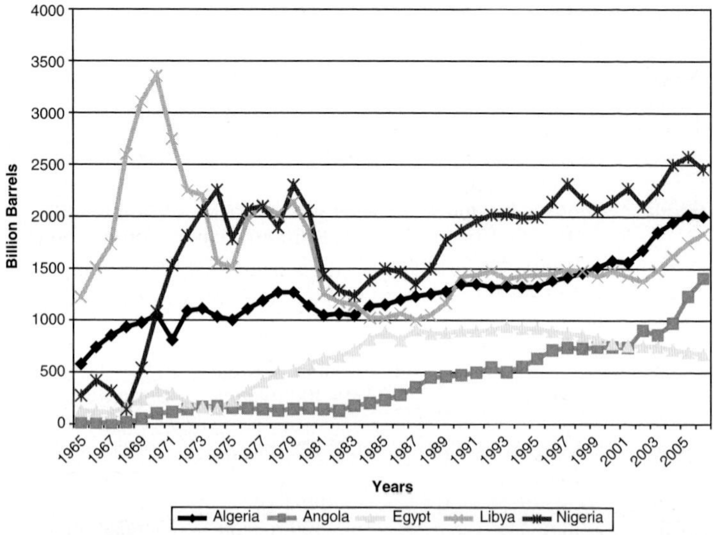

Figure 2.3: Oil Production in Africa by Country, 1965–2006

Source: Authors, with data from *BP Statistical Review of World Energy* (BP, 2007).

production rose from 1,084 bpd in 1970 to 2,256 bpd in 1974. In 1982, African production fell to its minimum of 4,814 bpd following the world economic recession. However, in the last decade virtually all producing African countries have registered a steady increase in production.

Oil production in Africa started in earnest in the 1960s and has been increasing gradually ever since, except for a slowdown in the early 1980s owing to the collapse in oil prices. Figure 2.4 shows this trend, from 2.2 million bpd produced in 1965 to about 10 million bpd in 2006.

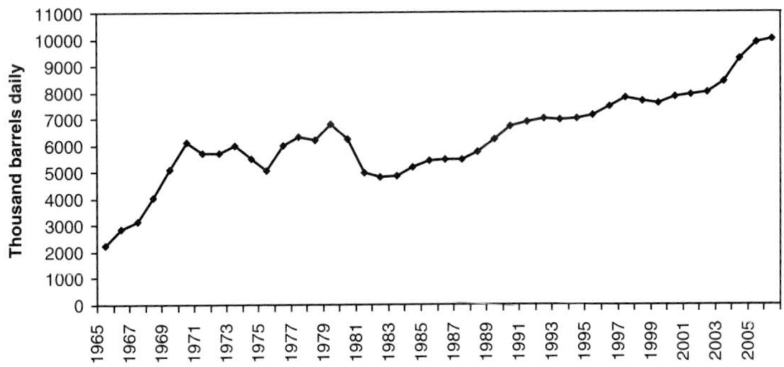

Figure 2.4: Africa's Average Oil Production per Year (Daily Production, bpd)

Data Source: *BP Statistical Review of World Energy* (BP, 2007).

2.3.3 *Exploration and Planned Production*

Oil exploration activities are generally led by oil companies. It is rare for governments to directly lead these activities, although nationalized or state-run oil companies are often involved. Oil exploration efforts in Africa have been boosted significantly in recent years, as illustrated by the rate at which proved reserves are growing over time. Figure 2.5 presents an overview of cumulative known reserves in Africa and illustrates the close correlation with the gradual rise in oil prices over the last 10 years. It is worth noting, however, that exploration, as well as estimates of reserves, increased gradually even in the 1980s, when oil prices were declining—most likely illustrating that exploration in Africa was underpinned by geopolitical and oil supply security motives (mostly from Western nations, in the early phase, but, increasingly, with emerging countries such as China).

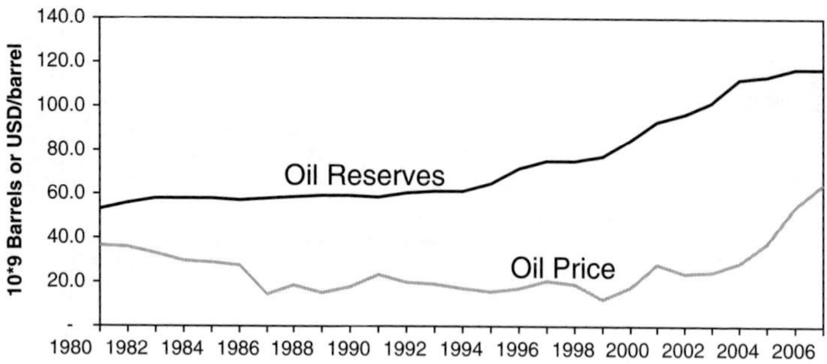

Figure 2.5: Proved Oil Reserves in Africa, Compared to Oil Prices (Yearly Average)

Data Source: *BP Statistical Review of World Energy* (BP, 2007).

Conventional wisdom holds that oil exploration increases when oil prices rise. This is attributable to the availability of large profits that companies can channel to exploration activities. At present, countries such as Ghana, Kenya, South Africa, Mozambique, Tanzania, Uganda, and several other African nations are being explored for hydrocarbonates, offshore and onshore. To date, large deposits of natural gas have been identified in Tanzania, and substantial probable oil deposits found in Albertine Graben (Uganda) and in the western part of Ghana (offshore). There are also potential significant oil discoveries in South Africa, Mozambique, and Tanzania.

2.3.4 *Main Trends and Expected Upstream and Downstream Developments*

The upstream industry comprises exploration and production activities, while the downstream industry

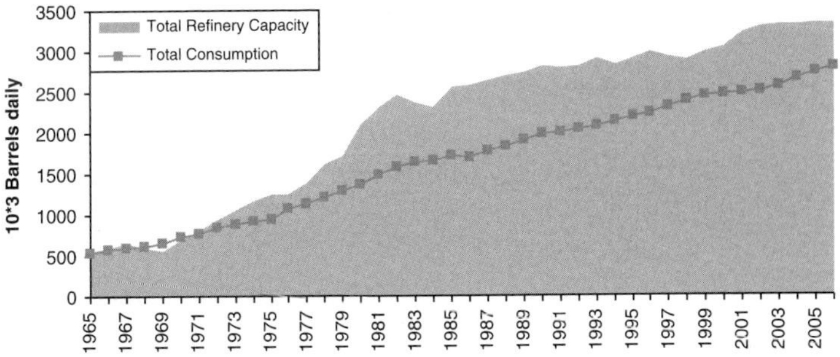

Figure 2.6: Total Oil Consumption and Refinery Capacity in Africa

Data Source: *BP Statistical Review of World Energy* (BP, 2006).

includes refining and retail activities. Oil refineries convert crude oil into fuel products, lubricating oils, bitumen, chemical feedstocks and other oil products (see Section 2.2).

The first African refineries were built in Algiers (CFP/Total) and Durban (Socony/Mobil) in 1954. In the 50 years up to 2004, a total of 48 refineries were built in Africa. However, the majority of them were established in the 1970s and 1980s. The only new refineries built in the last 10 years were Khartoum (Sudan) in 2001 and MIDOR (Egypt) in 2002. Africa's historic refinery capacity is presented in Figure 3.5. The continent's total active distillation capacity is about 3 million bpd (15 million mt/yr), an average of 79,000 bpd per refinery (Figure 2.6).

The major refining centers in Africa are located in South Africa (4 refineries and 3 synthetic fuel plants), Nigeria (3 refineries), Egypt (9 refineries), and Algeria (4 refineries) (Figure 2.7). The largest single refinery is

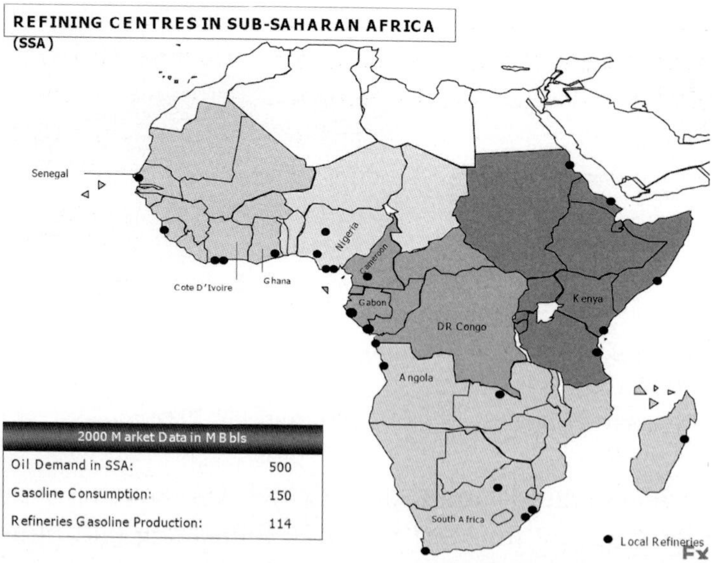

Figure 2.7: Refining Centers in Sub-Saharan Africa
Source: African Energy (2008).

the Skikda refinery in Algeria (300 million bpd), whereas the smallest operating refinery is the Solimar refinery in Madagascar, with a capacity of 14 million bpd (African Energy, 2008).

Many African refineries have been forced to close because of low worldwide refining margins, small local markets, high operating costs (due to small size), and poor yields. Following the World Bank/IMF insistence on market liberalization in the early 1980s, many of the remaining refineries have faced significant challenges. Although installed capacity in Africa is higher than present consumption (Figure 2.6), the continent still faces high shortages in refined products that are balanced by imports. As a consequence of such shortages and of the need to maximize economic profit by placing the refinery close to the source, several initiatives and plans are underway to install new refineries in such countries as Nigeria, Sudan, Uganda, and Mozambique, as outlined in Boxes 2.1 to 2.4.

Box 2.1: THE EXAMPLE OF NIGERIA'S REFINERY INDUSTRY

There are considerable investment opportunities in Nigeria's downstream oil and gas sector. The government's focus is on deregulating the sector by licensing private refineries, eliminating government subsidies to the downstream sector, and privatizing existing ones. Through such strategic action, domestic capacity is expected to at least meet demand. The four existing refineries have significantly and consistently produced below capacity, owing to a host of factors, including poor management and maintenance.

The deregulation of the downstream oil and gas sector (Petroleum Refining & Marketing) has been in focus for a number of years. Indeed, scarcity of petroleum products and gradual deregulation of petroleum product prices have generated heated controversies in Nigeria. The government is determined to nurture private-sector participation and engage local companies in the oil and gas sector—hence the licensing of private refineries and deregulation of petroleum product prices—to improve local capacity. In the coming years, consolidation is expected in the oil refining and marketing sector, with new entrants. There are significant investment opportunities and a need for both local and international funding.

Sahara Petroleum Exploration, a subsidiary of Global Environmental Energy, has been contracted to build a 70,000 bpd oil refinery at Eket, Akwa Ibom state, Nigeria.

This is an approximately $4 billion project that represents cash flow at today's market rate of approximately $1.5 billion per annum. Such plans to increase Nigeria's refining capacity will also provide considerable cost savings to operators (such as Sahara Petroleum) by refining oil products close to source. At present, the country has four main refineries with a nameplate capacity of 438,750 bpd.

Data Sources: Market Research (2008) and Onlinenigeria.com (2005).

Box 2.2: TULLOW OIL REFINERY PLAN IN UGANDA

Tullow Oil is currently working to find out the quantity and quality of oil that has been discovered in the Albertine Graben in Uganda before a decision is made to build a multi-billion dollar refinery or not. The Uganda government and Tullow Oil have agreed to build a mini refinery in an early production scheme that will enable installation of a 100 MW heavy fuel oil thermal power plant. The mini refinery (with a capacity of about 5,000 barrels of oil per day) will cost Tullow some US$200 million.

Data Source: Kisambira (2007).

Box 2.3: SUDAN TO DOUBLE ITS OIL-REFINING CAPACITY

Sudan is planning to double its oil refining capacity, with a short, three-year investment plan, to handle increased production, having reached an agreement to end its 21-year civil war. A new 100,000 bpd refinery will be built in Port Sudan, on the Red Sea coast, and the capacity of two existing refineries will be increased. Talks have been held with India's Oil & Natural Gas, China's Sinopec, Malaysia's Petronas, and an unidentified Turkish company to build the new refinery. Boosting refinery capacity is especially timely since Sudan expects crude oil production to increase significantly in the coming years.

Data Source: Chmaytelli (2005).

Box 2.4: NEW MOZAMBIQUE OIL REFINERY APPROVED

The Mozambican Council of Ministers has approved the construction of an oil refinery, valued at more than $1.3 billion, in the northern province of Nampula. Dubbed the 'Ayr Logistics Limited—Nacala', the project is spearheaded by a privately owned American company, Ayr Logistics, in partnership with one Mozambican and three South African investors. The project is expected to create about 450 jobs and generate extra tax revenue for the Mozambican government. With an installed capacity of about 100,000 bbl/d, most of the product will be exported to Malawi, Zimbabwe, and Zambia. The project also includes the construction of several infrastructures that will support the main activity of the project, which will be implemented over an area of 838 ha and will be situated in the district of the Nacala port, which is also home to Mozambique's deepest and busiest port.

Data Source: *Engineering News* (2007).

Despite the plans and visions illustrated in the boxes, the fact remains that very little new global refining capacity has been added in the last three years, including in 2007. Nevertheless, significant refinery capacity additions are *still* planned, although a major concern is that construction costs are rising with inflationary prices. Other non-economic reasons—important in Africa, as elsewhere—for the slow growth in refining capacity (worldwide) are environmental and local concerns, more stringent environmental laws, and effective community organizing, which have made it very difficult to build new refineries. Some analysts consider refinery capacity a significant factor of high oil prices. It thus makes sense to increase refinery capacity in Africa as a way to reduce the costs of refined oil products. However, this needs to be done in an environment-friendly manner.

2.4 Oil Trade in Africa

The oil market can be divided into two parts: within Africa and between Africa and the rest of the world. The major trends are clear: There is significant oil trade within West Africa and North Africa, whereas there is very little oil trade between African producers and the

Table 2.9: Exports of Fuels and Mining Products from Africa, 2006 (Billion Dollars and Percentage)

	2006	2000	2006	2000	2006	2000–2006	2005	2006
Africa								
World	249.0	100.0	100.0	10.0	10.9	19	44	26
Europe	92.3	44.6	37.1	4.5	4.1	16	47	23
North America	69.7	24.2	28.0	2.4	3.1	22	55	26
Asia	54.6	20.0	21.9	2.0	2.4	21	35	34
Africa	13.0	5.1	5.2	0.5	0.6	20	36	33
South and Central America	10.2	4.1	4.1	0.4	0.4	19	34	24
Middle East	1.4	0.9	0.5	0.1	0.1	10	17	19
Commonwealth of Independent States (CIS)	0.1	0.0	0.1	0.0	0.0	23	53	8

Source: WTO (2007), International Trade Statistics, WTO, Geneva.

Eastern and Southern Africa sub-regions. Table 2.9 shows exports of fuel and mining products from Africa to various regions of the world. In 2006, for example, exports from African producers to African consumers represented a meager 5.2 percent (almost unchanged from the 5.1 percent recorded in 2000) while the largest share of 37.1 percent went to Europe, followed by 28 percent to North America, and 21.9 percent to Asia.

The energy balance for Africa is presented in Table 2.10, which shows that significantly larger amounts are exported than imported. This is good from a trade and economic standpoint. However, it is worth noting that even net oil producers import oil derivatives and products.

Table 2.11 presents oil imports and exports for Africa by region (2007 data). The data clearly shows Africa's

Table 2.10: Energy Balance for Africa (in Thousand Tons of Oil Equivalent, 2006 (KTOE))

Supply and Consumption	Coal and Peat	Crude Oil	Petroleum Products	Gas	Nuclear	Hydro	Geothermal, Solar, etc.	Combustible Renewables and Waste	Electricity	Heat	Total*
Production	141,801	495,846	0	169,668	3,070	8,235	939	29, 0921	0	7	1,110,488
Imports	7,698	42,879	45,214	4,030	0	0	0		2,734	0	102,556
Exports	−46,879	−401,357	−43,524	−96,473	0	0	0	-282	−2,742	0	−591,258
International Marine Bunkers**	0	0	−6,035	0	0	0	0	0	0	0	−6,035
Stock Changes	−39	−336	−1,113	0	0	0	0	0	0	0	−1,489
TPES	102,581	137,032	−5,458	77,224	3,070	8,235	939	290, 640	−8	7	614,262
Transfers	0	−12,088	12,832	0	0	0	0	0	0	0	744
Statistical Differences	−414	−313	166	110	0	0	0	0	172	0	−278
Electricity Plants	−61,712	−765	−15,808	−37,460	−3,070	−8,235	−939	-566	50,543	−7	−78,020
Gas Works	−4,108	0	0	1,876	0	0	0	0	0	0	−2,232
Petroleum Refineries	0	−130,454	126,728	0	0	0	0	0	0	0	−3,726
Coal Transformation	−2,291	0	0	0	0	0	0	0	0	0	−2,291
Liquefaction Plants	−17,168	7,410	0	−3,794	0	0	0	0	0	0	−13,552
Other Transformation	0	0	0	0	0	0	0	-35, 546	0	0	−35,546
Own Use	−12	−639	−4,528	−12,067	0	0	0	0	−3,880	0	−21,127
Distribution Losses	−117	−183	0	−1,128	0	0	0	0	−5,776	0	−7,204
TFC	16,758	0	113,932	24,762	0	0	0	254, 528	41,050	0	451,030

Source: IEA (2008).

* Totals may not add up due to rounding.

** International marine bunkers are not subtracted out of the total primary energy supply for world totals.

NB: TPES = Total Primary Energy Supply; TFC = Total Final Consumption.

Table 2.11: Oil Imports and Exports for Africa by Region, 2006

	Thousand barrels per day			
	Crude Imports	Product* Imports	Crude Exports	Product* Exports
North Africa	179	178	2,721	615
West Africa	68	234	4,706	123
East & Southern Africa	514	176	385	22
TOTAL WORLD	39,836	14,988	39,836	14,988

**Products*: Petroleum product derivatives.
Data Source: BP (2008).

Table 2.12: Net Oil Exports for Countries in Africa

Year	2000	2004	2005	2006
Country	(Thousand barrels per day)			
Algeria	1,276.86	1,733.79	1,840.34	1,846.73
Angola	716.86	1,003.48	1,206.70	1,363.23
Cameroon	62.33	42.21	58.52	62.59
Chad	−1.34	169.13	175.16	156.37
Congo (Brazzaville)	275.38	229.16	222.09	240.89
Congo (Kinshasa)	11.86	12.89	12.88	14.25
Cote d'Ivoire	−20.53	12.85	34.49	67.97
Egypt	279.13	112.70	95.20	68.61
Equatorial Guinea	165.97	370.47	354.39	334.64
Gabon	302.65	225.80	253.56	224.94
Libya	1,259.08	1,327.60	1,452.41	1,525.49
Mauritania	−23.86	−24.20	−24.21	6.40
Nigeria	1,923.59	2,042.56	2,336.95	2,146.31
Sudan	143.88	278.66	273.58	279.27
Africa	6,371.84	7,537.10	8,292.06	8,337.69

*Oil includes crude oil, lease condensates, natural gas liquids, other liquids, and refinery gain.
Data Source: EIA (2006).

limited significance with respect to imports as well as to derived products, but its very significant contribution to basic world crude oil exports. Table 2.12 presents export data from 2000 to 2006 for the 14 African net oil-producing countries.

It should be noted that only two of the 14 net oil-exporting African countries are *landlocked* (Chad and Mauritania). Marketing (exporting) crude oil from a landlocked country is an added challenge for an investor. The case of Chad (Box 2.5) illustrates an effort to solve this problem.

Box 2.5: CHAD–CAMEROON PIPELINE

Chad has a long history of civil war and exhibits many conditions associated with post-conflict zones. To overcome its geographic disadvantage of being landlocked, Chad needed a pipeline to exploit its over one billion barrels in proven oil reserves. In 1999, conditions inside Chad were so bad that no one in the private oil sector was willing to invest in a pipeline unless the World Bank was involved. In the first project of its kind, the World Bank agreed to provide investment coupled with institutional oversight and transparency (U.S. Department of Energy, 2007*b*). The World Bank and a consortium of oil companies, led by ExxonMobil, ChevronTexaco, and Petronas, set up a pipeline project in Chad.

Construction of the $3.5 billion Chad–Cameroon Petroleum Development and Pipeline Project began in 2000 and was completed in 2004. To increase transparency, Chad was required to adopt the Petroleum Revenues Management Law, which stipulated that Chad's 12.5 percent of the oil revenues would be deposited into a Citibank escrow account monitored by an independent "college" before the Chadian government received it. Another 10 percent was deposited in a "future generations" fund to provide Chad with revenues after the exhaustion of the oil reserves (Zissis, 2007). However, in December 2005, Chad's National Assembly abolished the future generations fund and diverted money away from poverty-mitigation efforts to arms purchase. The World Bank responded by suspending $124 million in loans.

(*cont.*)

Box 2.5: (*Continued*)

In July 2006, the two sides reached a compromise, which specified that the Chadian government would commit 70 percent of revenues to development programs and 30 percent to government expenditures (Thibodeaux, 2007). Some have called the Chad experiment a *prima facie* failure; others maintain that after only a few years it is still too early to judge.

2.5 Africa's Consumption of Oil

Figure 2.8 shows the growing pattern of oil product type consumption in Africa. Consumption of oil derivates in Africa has grown steadily, with total consumption in 2006 close to 3 million barrels per day. However, this only represents about 3 percent of world consumption.

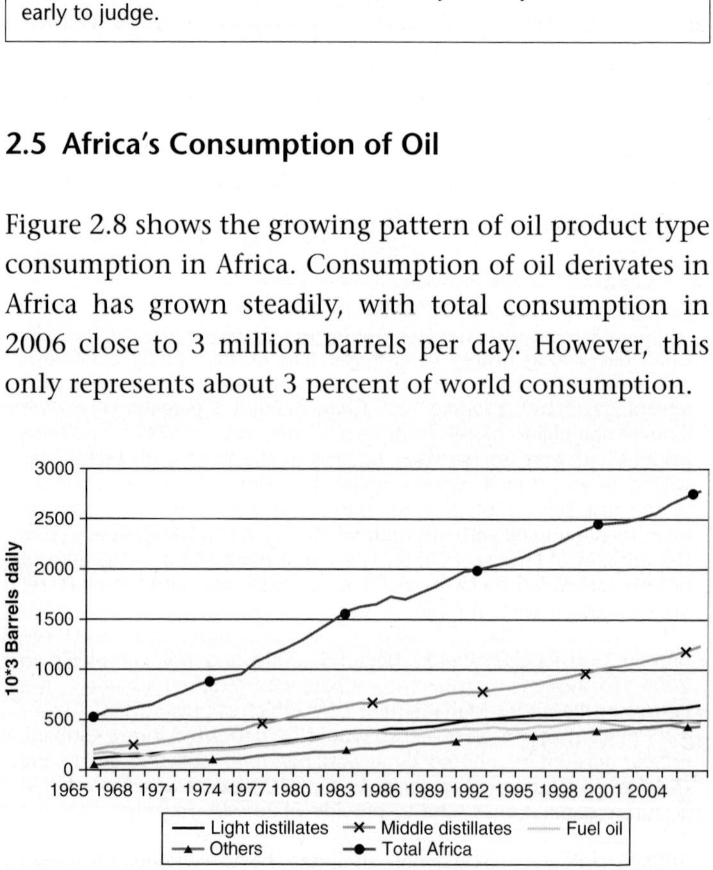

Figure 2.8: Africa's Oil Consumption by Product Type (barrels per day)

Data Source: BP (2008).

Table 2.13: World and Africa Total Proved Natural Gas Reserves, 1986–2006

	At end 1986 (trillion cubic meters)	At end 1996 (trillion cubic meters)	At end 2005 (trillion cubic meters)	At end 2006 (trillion cubic meters)
Algeria	3.26	3.70	4.50	4.50
Egypt	0.29	0.85	1.90	1.94
Libya	0.73	1.31	1.32	1.32
Nigeria	2.40	3.48	5.15	5.21
Other Africa	0.72	0.83	1.21	1.21
Total Africa	7.40	10.17	14.08	14.18
TOTAL WORLD	107.67	147.89	180.20	181.46

Data Source: BP (2007).

2.6 Gas Reserves

Historically, world natural gas reserves have trended upward, for the most part. As of January 1, 2007, world proved natural gas reserves, as reported in *Oil & Gas Journal* (2007), were estimated at 6,183 trillion cubic feet. World proved reserves of natural gas grew 68.5 percent between 1986 and 2006. Over the same period, proved natural gas reserves in Africa grew 91.6 percent to a world share of 7.8 percent. Table 2.13 shows the growth pattern of proved natural gas reserves in Africa and in the world. Almost three-quarters of the world's natural gas reserves are located in the Middle East and Eurasia. Russia, Iran, and Qatar, combined, accounted for about 58 percent of the world's natural gas reserves as of January 1, 2007. Reserves in the rest of the world are fairly evenly distributed on a regional basis (EIA, 2007).

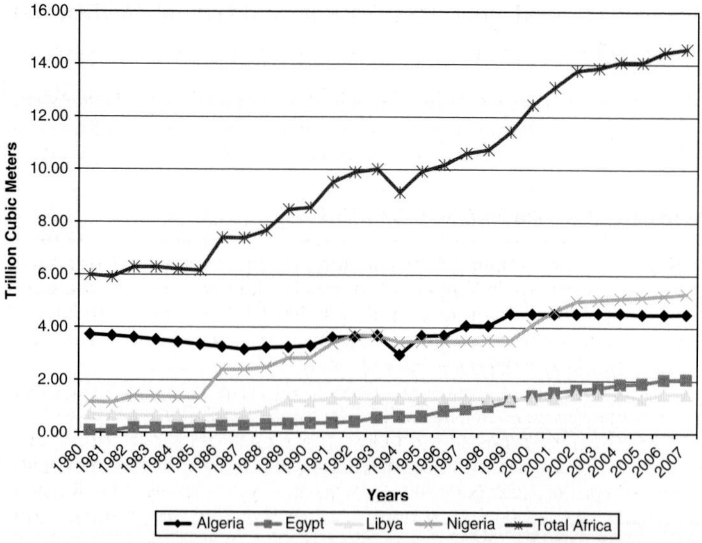

Figure 2.9: Africa's Proved Natural Gas Reserves, 1980 and 2007

Data Source: BP (2008).

Despite the high rates of increase in natural gas consumption, particularly over the past decade, most regional reserves-to-production ratios are substantial. Worldwide, the reserves-to-production ratio is estimated at 65 years. Central and South America have a reserves-to-production ratio of about 52 years; Russia, 80 years; and Africa, 88 years. The reserves-to-production ratio of the Middle East exceeds 100 years (EIA, 2007).

Proved natural gas reserves in Africa are concentrated mainly in four countries—Algeria, Egypt, Libya, and Nigeria—which possess 91.5 percent of proved reserves. Figure 2.9 presents the distribution pattern of proved natural gas reserves per country in Africa. The undeveloped natural gas reserves in Nigeria, in particular, are a target of the international giants in the sector (Box 2.6).

Box 2.6: NIGERIA NATURAL GAS AT STAKE

Nigeria exports about 18 million tons of liquefied natural gas each year through Nigeria Liquefied Natural Gas, jointly owned by the state energy company, Shell, Total, and Agip. Nigeria has the seventh-largest proven gas reserves in the world, but it has not developed its gas industry to anywhere near full potential.

According to investors, most investment ideas in the sector are not profitable because of the lack of a stable fiscal framework and of market pricing for gas. Nevertheless, the Russian gas export monopoly, Gazprom, is in talks with Nigeria to spend up to $2.5 billion to exploit, gather and process Nigeria's vast natural gas reserves. The Russian company is reported to be on a global hunt for new reserves and has emerged as one of the leading players in the Nigerian natural gas sector. UK-based companies BG Group and Centrica have also proposed multi-billion dollar investments.

About 2.5 billion cubic feet of Nigerian gas associated with the extraction of crude oil is burnt off every day because there is no infrastructure to use the gas. The government is planning to fine companies that flare gas. It hopes to see an explosion in domestic demand for gas for use in power generation, petrochemicals, and fertilizers. But the price the state power monopoly now pays for gas makes investment in stand-alone gas projects unprofitable.

Source: Adapted from Fin24 (2008) (Fin24.com article).

2.7 Gas Production

Natural gas production in Africa rose from 2.8 billion cubic meters (BCM) in 1970 to 50.2 BCM in 1986 and 190.4 BCM in 2007, representing about 6 percent share of world production. Production capacity in Africa grew

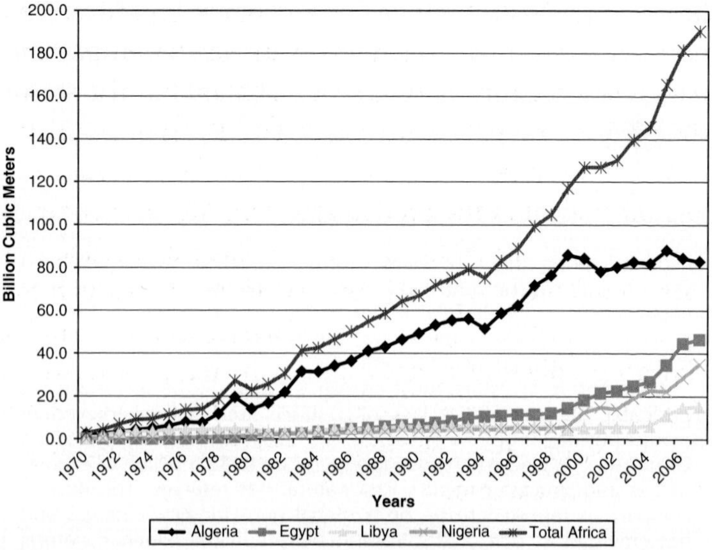

Figure 2.10: Africa's Natural Gas Production, 1970 and 2007
Data Source: BP (2008).

by 359.5 percent between 1986 and 2006. Algeria is the leading African producer of natural gas, representing 2.9 percent of the total world share, followed by Egypt and Nigeria (Figure 2.10). Map 2.1 presents the location of known major deposits on the continent.

According to EIA (2007), Africa and non-OECD Asia (excluding China and India) are expected to be important sources of natural gas production in the future. For each of the two regions, natural gas production in 2030 is projected to be some 10 trillion cubic feet above 2004 production levels. The two regions combined accounted for 14 percent of the world's natural gas production in 2004; in 2030, their combined share is projected to be 21 percent. A substantial portion of the production from both regions is exported. In 2004, 26 percent of natural gas production from non-OECD Asian countries (primarily Brunei, Indonesia, Malaysia, and Myanmar) and 50 percent of the production from African countries was exported. In 2030, the export share of production from non-OECD Asia is projected to fall to 10 percent as domestic consumption rises, while the export share of Africa's production is projected to increase even faster. Several pipelines from North Africa to Europe are under consideration, and LNG export capacity in West Africa continues to expand.

2.8 Consumption of Gas

Worldwide consumption of natural gas will increase from 100 trillion cubic feet in 2004 to 163 trillion cubic feet in 2030 in the *EIA 2007* reference case. By energy source, the projected increase in natural gas consumption is second only to coal. Natural gas remains a key fuel in the electric power and industrial sectors. In the power sector, natural gas is an attractive choice for new generating plants because of its relative fuel efficiency. Natural gas also burns more cleanly than coal or petroleum products. As more governments begin implementing national or regional plans to reduce carbon dioxide emissions, they may encourage the use of natural gas to displace liquids and coal.

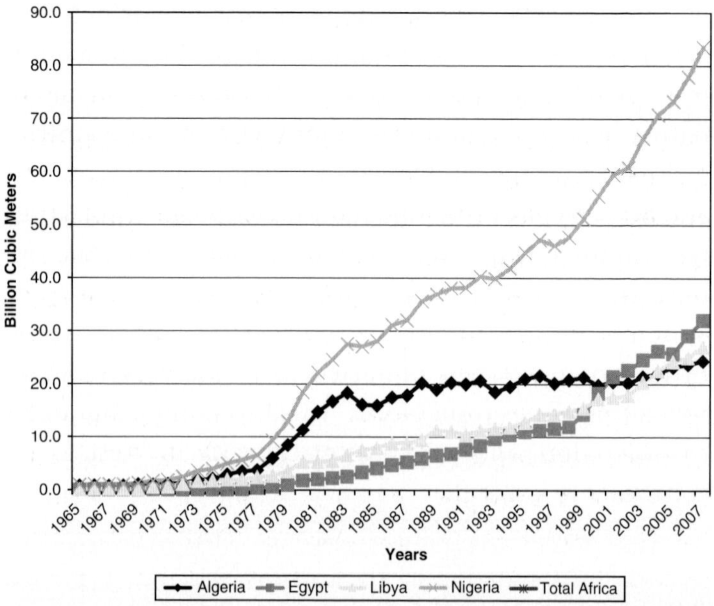

Figure 2.11: Africa's Natural Gas Consumption, 1965 and 2007 (BCU)

Data Source: BP (2008).

Consumption of natural gas in Africa is still very low. Only countries like Egypt, Algeria, and South Africa have relatively high consumption levels (Figure 2.11). This is mainly associated with previous investments and with the high economic performance of these countries. Africa's natural gas consumption in 1965 was 1 BCM; this increased to 31.1 BCM in 1986 and 75.8 BCM in 2006, representing about 2.6 percent of current world consumption.

2.9 Natural Gas Market Dynamics

Natural gas prices were steady and low until 2000, when they began to follow world oil prices, rising sharply from 2 to 9 USD per million BTU for CIF Europe (Figure 2.12).

The dynamics of natural gas markets are fundamentally different from those of crude oil, which is extremely sensitive to geopolitical events. Natural gas is bought and sold through a large, integrated, continental scale market with multiple pricing hubs. Natural gas transfers on the TransCanada Alberta System represent the main Canadian pricing hub (AECO "C"), while the Henry Hub in Louisiana is the pricing point for natural gas traded on the New York Mercantile Exchange (NYMEX). The price of natural gas traded at these hubs is publicly posted and sets a commodity cost for natural gas. Natural gas

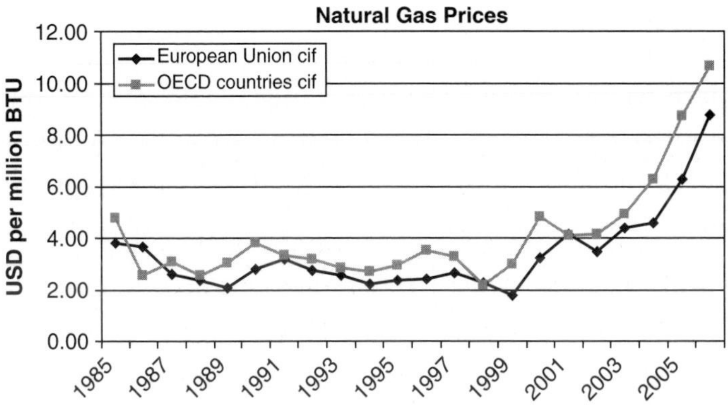

Figure 2.12: Natural Gas Prices (CIF Europe and OECD Countries)

Data Source: BP (2007).

Box 2.7: NIGERIA ALONE COULD MEET THE POWER NEEDS OF ALL OF WEST AFRICA

Nigeria is gradually trying to develop a regional market for its gas. The government is working on a plan to build about 600 km of new gas pipeline to Benin, Togo, and Ghana. The governments of these four West African nations agreed in 1995 to build the pipeline, which, with the existing pipeline, would stretch about 960 km. This pipeline project, according to World Bank studies, could save West African countries importing the gas about $500 billion in primary energy costs over 20 years.

The pipeline will provide markets for Nigeria's natural gas, thus earning hard currency for the country's economic and social development. It will also enable a reduction in gas flaring, leading to a safer environment in Nigeria, and contribute to the long-desired economic integration of the region.

Nigeria is potentially capable of fueling the power needs of the whole of West Africa.

Adapted from: Obadina (2000).

prices are highly influenced by contracts between natural gas buyers and sellers trading in the natural gas futures market. Futures contracts quote prices for delivery of a specified quantity of natural gas at a particular time and place in the future. Delivery ranges from one or two months from the date of inception to several years in the future. The futures contracts are traded exclusively on regulated exchanges and are settled daily based on their current value in the marketplace (CNGMO, 2007).

The price consumers pay is the sum of three parts: the commodity cost of the natural gas, the pipeline transportation cost, and the distribution cost (CNGMO, 2007). The trade movement of liquefied natural gas in Africa is geared to the international markets, the largest importers being Turkey, Italy, Greece, and

Table 2.14: Natural Gas: Trade Movements, 2006—Liquefied Natural Gas (LNG)

To	From (in BCM)				(BCM) Total Imports
	Algeria	Egypt	Libya	Nigeria	
North America					
United States	0.49	3.60		1.62	16.56
Mexico		0.16		0.54	0.94
Europe					
Belgium	3.35	0.25		0.16	4.28
France	7.35	2.30		4.23	13.88
Greece	0.45	0.04			0.49
Italy	3.00	0.10			3.10
Portugal				1.97	1.97
Spain	2.80	4.80	0.72	7.10	24.42
Turkey	4.60			1.12	5.72
UK	2.00	0.96			3.56
Asia Pacific					
China					1.00
India	0.08	0.55		0.08	7.99
Japan	0.24	0.80		0.22	81.86
South Korea	0.32	1.25		0.16	34.14
Taiwan		0.16		0.38	10.20
TOTAL	24.68	14.97	0.72	17.58	210.11

Data Source: *BP Statistical Review of World Energy* (BP, 2006).

France, which import nearly 100 percent of their needs from Africa. The largest exporter in Africa is Algeria, which exports 24.68 BCM per year (Table 2.14).

2.10 Africa's Position on the Global Energy Scene

Most African oil producers are members of OPEC (Organization of the Petroleum Exporting Countries), whose stated goal is to promote sovereign control over resources and attempt to stabilize world oil markets with fair prices for producers and guaranteed security of supply to consumers.

Considering the current energy uncertainties, the key drivers of future demand, consumer countries' policies (especially on nuclear and other alternatives to oil and gas), and expected future global economic growth and technology development, there is need to clearly establish Africa's position and develop strategies for future supply adequacy.

The evidence indicates that gas is gaining a significant share in the global energy mix. Indeed, the World Bank estimates that the global demand for gas will outstrip that for oil as early as 2025. This projected dominance of gas is driven by an assortment of factors that include accelerated growth in world economic activity, changing energy markets, and advances in technology (for example, gas-to-liquids and compressed natural gas)—all of which have combined to transform gas into a cost-competitive and flexible global energy option at a pace few thought possible. In addition, gas offers significant environmental advantages over oil and coal and, as such, has become a preferred fuel in the power sector. At the same time, indigenous gas production in Europe and America is on the decline, with nearly half of total supplies to the United States and Europe projected to be imported by 2015.

Africa's rich oil fields and the prospects for more discoveries have transformed the continent into an important player in global oil production. The continent holds approximately 9.7 percent (Table 2.1) of the world's oil reserves and accounts for 12.1 percent of world oil production (2006). Africa's oil production is expected to continue to rise at an average rate of 6 percent per year for the foreseeable future. There is evidence from several sources of Africa's growing importance within global energy markets: This includes recent large deepwater oil discoveries in Angola, Nigeria, and Equatorial Guinea. The waters around São Tomé and Principe are believed to hold about two billion barrels of oil reserves and the strong interest from international oil companies suggests that this tiny archipelago will join a growing group of African oil producers. Thanks to the opening of the Chad–Cameroon pipeline in July 2003, Chad's oil reserves can find their way to global markets (see Box 2.5). Oil has also recently been discovered in Niger, Ghana, and Uganda. Unsurprisingly, Africa is now looked upon as a key source for major oil importers seeking to diversify imports in order to achieve greater energy security (Malaquias, 2005).

Despite the major political and economic challenges confronting Africa, international oil companies have continued to invest massive amounts of capital in further exploration and development of the region's oil fields for a variety of reasons. First, foreign investors are attracted by very competitive terms and conditions offered by African governments. Second, much of the oil is explored offshore (but not in deep waters) and therefore has the advantage of being easily loaded and moved anywhere by ship. Also, being offshore, African oil is located far away from onshore sites of disorder. Third, the relative proximity to the American and European markets and the fact that African oil tends to be particularly suited for American refineries because of its high quality and low sulfur provides an added incentive, especially for U.S. investors (Box 2.8).

Box 2.8: AFRICA'S OIL EXPORT TO THE UNITED STATES

African oil production has risen constantly over the last few years. Indeed, African countries currently supply about 15 percent of U.S. oil imports. Nigeria produces about 2.12 million b/d and exports 1.85 million b/d, of which 621,000 b/d to the United States, making it the United States' fifth largest supplier. Much of Nigeria's crude oil production—about 65 percent—is light and sweet. This makes it particularly suited for American refineries because it yields high volumes of gasoline. Nigeria has a bright energy outlook and is expected to significantly increase its crude oil production in the next few years as recent deep-water discoveries come on stream. Angola produces 900,000 b/d and exports 866,000 b/d (332,000 b/d to the United States, making it the United States' ninth largest supplier, and third largest non-OPEC supplier outside of the western hemisphere). Angola also has a very promising future energy outlook, especially as oil begins to flow from large fields in the Kwanza basin south of Cabinda. Cameroon, Chad, Equatorial Guinea, and Gabon export approximately 500,000 b/d in aggregate (221,000 b/d to the United States).

Data Source: Malaquias (2005).

For China, Africa has become an important area on the global energy scene. China currently derives a quarter of its oil imports from Africa; it holds significant oil interests in Algeria, Angola, Chad, and Sudan (Boxes 2.9 and 2.10) and increasing stakes in Equatorial Guinea, Gabon,

Nigeria, and even Chad—a country that has diplomatic relations with Taiwan. Chinese firms are becoming increasingly involved in the Nigerian oil sector.

Box 2.9: CHINA—ENGAGING IN OIL EXTRACTION IN SUDAN

China first successfully established an oil outpost in Sudan in 1995 when its National Petroleum Corporation gained oil exploration rights there. This presence grew when, in 1997, the United States imposed comprehensive economic, trade, and financial sanctions against Sudan. China quickly filled the vacuum.

Chinese investment and technical expertise played an important role in establishing Sudan's oil industry, which began producing oil in 1999. Sudan currently produces about 500,000 b/d.

China's National Petroleum Corporation is the largest shareholder in the consortium of international oil companies controlling much of Sudan's energy sector. More than half of Sudan's oil exports go to China, accounting for 5 percent of China's total oil imports. China has also become the biggest investor in Sudan's $15 billion, 932-mile oil pipeline to Port Sudan on the Red Sea, where China has built a tanker terminal.

Data Source: Malaquias (2005).

Box 2.10: CHINA—ENGAGING IN ANGOLA'S OIL SECTOR

China's growing influence in Angola's oil sector has been facilitated by a series of soft loans, including a recent agreement whereby 10,000 bpd of crude oil was tied to a US$2 billion loan to be repaid over 17 years at 1.5 percent interest. These are much better terms than Angola could ever hope to secure from other sources. This loan, part of a larger aid package, represents a significant departure from the traditional way Angola has engaged external players—mostly through deals involving production-sharing agreements, market contracts, or international finance agreements.

Data Source: Malaquias (2005).

Africa is endowed with vast quantities of both fossil and renewable energy resources. It is the only continent in the world with frequent and substantial new finds of oil and gas. Thus, in the last 20 years, oil reserves have grown by over 25 percent, while gas reserves have grown even faster, by over 100 percent. Exploiting these resources in a sustainable manner for the benefit of the populations is crucial for Africa's future development.

Oil and gas have until now been concentrated in pockets in North, West, and Eastern Africa. Coal reserves are dominant in the southern region, with over 95 percent in South Africa alone. As a result of this skewed distribution, more than 70 percent of countries in the continent import oil and gas to satisfy their needs (Davidson and Sokona, 2002).

Geothermal deposits are available in the continent, but are mainly limited to Eastern Africa, along the Rift Valley, and concentrated in Kenya and Ethiopia. In addition, Africa has over 10 percent of world hydro resources and significant other renewable energy resources. Most countries in the continent enjoy long hours of sunshine with significant radiation that can be exploited. Wind resources are available in selected sites, mostly along the coastlines in Northern, Western, and Southern Africa.

In principle, therefore, Africa is not short of energy resources, which can give the continent the energy security it requires. The challenge is to develop strategies and invest adequately—for example, access well-known technologies and overcome the major barriers that exist. Energy *insecurity* undermines human development and institutional capacities and lowers economic growth. Energy security is correspondingly closely linked with social, economic, political, and environmental development, thus making it a cross-cutting issue rather than one of simply finding geological energy resources.

2.10.1 *Africa's Imminent Energy Crisis*

Africa currently faces an energy crisis. As late as January 2008, South Africa, for instance, faced its worst energy crisis, believed to be the result of inadequate forward-planning for energy supply relative to economic growth (and corresponding energy demand) in the country and in the region.

According to Schultz (2007), taxi fares in Dakar, Senegal's capital city, have almost doubled since 2005 and blackouts occurred frequently in 2006 because state-owned utilities could not afford to pay for fuel. The country relies on oil imports to power its diesel-fired generators. Although conditions have improved somewhat, power cuts are on the rise again. As of May 2007, the capital was facing 10-hour power cuts several times a week and the government was warning of impending, unprecedented shortages. Senegal is paying nearly twice what it was a few years ago to import the same amount of oil. The increased cost alone is more than seven times as much as the country is gaining through multilateral debt relief programs. The government has responded to the energy crisis by providing direct subsidies to consumers. Since the rise in world oil prices began in 2002, these subsidies have increased five-fold, creating yet another burden on the national budget.

Uganda has been plagued by increasingly frequent and severe power outages from its hydroelectric stations on the upper tributaries of the Nile River (Lake Victoria). Due in part to global warming, the water levels of Lake Victoria, the largest of Africa's great lakes, have been decreasing steadily over the last decade. As a result of regional droughts, water levels dropped an astonishing half an inch per day in most of 2006 (Schultz, 2007) and Uganda experienced one of the worst power shortages in its history. As a result, the government resorted to an extreme load-shedding regimen, providing power only every other day. Responding to the emergency with the support of the World Bank, it installed two diesel-fired generators to relieve the shortage—a less-than-perfect solution.

Similar trends are occurring elsewhere in Africa, where scarce budgetary resources, desperately needed in the health and education sectors, are being spent to cushion oil and electricity costs. It is estimated that poverty has increased as much as 6 percent in some parts of the world due to the hike in oil prices in recent years. Especially vulnerable are highly indebted countries, which rely on oil imports to fuel their economy. The poorest countries in the world consume an almost negligible share of the millions of barrels of oil consumed every day globally, yet they are hit the most by rising world oil prices—and then hit again by the effects of climate change associated with burning hydrocarbons. The solution to this crisis requires that countries develop the tools to diversify their energy sources away from conventional fossil fuels (see Schultz, 2007). These issues are further analyzed in Chapters 4, 5, and 6.

2.11 Summary and Conclusion

Energy has traditionally played an important role in all nations because it is an indispensable input for economic growth and development. Two-thirds of global energy requirements are met from oil and gas supplies. Evidence has shown that there is a strong correlation between energy consumption per capita and the level of economic and social progress.

The regional composition of global energy consumption reveals a wide disparity in global access to reliable and adequate commercial energy. Africa, with about 15 percent of the world's population, consumes 3 percent of global commercial energy. One can reasonably assume that Africa's low energy consumption level and share reflect both its low access to affordable commercial energy and low level of development and industrialization. Africa's share in global energy production is about 12 percent, and trending upwards. Its oil production is expected to continue to rise at an average rate of 6 percent per year for the foreseeable future. Evidence of the continent's growing importance within the global energy markets comes from several sources, including recent large deepwater oil discoveries in Angola, Nigeria, and Equatorial Guinea, as well as new discoveries in such countries as Uganda and Ghana.

Africa holds approximately 9.7 percent of world oil reserves (2006). Its rich oil fields and the prospects for more discoveries have transformed it into an important player in global oil production. However, its role is less significant in terms of import or consumption of oil and oil products (3–4 percent level), although consumption is on the rise with recent improved economic performance.

The proved natural gas reserves in Africa are concentrated mainly in four countries—Algeria, Egypt, Libya, and Nigeria—which possess 91.5 percent of the proved reserves and accounted for 7.8 percent of world proved reserves in 2006. Natural gas production and consumption in Africa are very low, representing 6.3 percent and 2.6 percent of world production and consumption, respectively.

World oil prices have been trending higher since 2000, and natural gas prices have tracked along. In recent years, several factors have been responsible for the current state of the oil market, highlighted by very volatile and rising prices of crude oil and petroleum products. Some of the reasons attributed to the rise in oil prices include rising demand in emerging economies, especially in China and in India; declining spare capacity in major producing countries; peaking of production in several important oil-producing areas; and lack of expansion in refinery capacity.

The issue of the *macroeconomic* effects of higher oil prices has been a recurrent theme since the first major oil price shock in the 1970s. Historical evidence shows that oil price increases and volatility have had a significant impact on global, regional, and national economic performance and outlook. Conventional wisdom, backed by empirical analysis, suggests that sustained and sharp increases in oil prices tend to exert strong inflationary pressures, reduce output, slow down economic growth, and exacerbate unemployment in net oil-importing countries—to which no fewer than 38 countries in Africa belong (analyzed further in Chapter 4).

The refinery industry in Africa lacks competitiveness and economies of scale. Most refineries are economically inefficient, operating below 50 percent capacity. There are different regional and country visions aimed at increasing capacity and improving performance, but, by the end of 2007, very little had happened.

In light of current energy uncertainties, key drivers of future demand, consumer countries' policies (especially on nuclear and other alternatives to oil and gas), and expected future global economic growth and technology development, there is need to clearly establish Africa's position and develop strategies for future supply adequacy. The energy crisis, which is exacerbated by the public outcry on the use of petroleum products—due to its high prices, environmental impacts, and growing concern about the viability of oil-based fuels and products—is leading to initiatives to find and develop alternative energy sources (Chapter 4 discusses in detail coping mechanisms for high oil prices).

3

Maximizing the Benefits from Africa's Oil and Gas Resources

3.1 Introduction

Africa is blessed with vast natural resource-rich environments. These resources constitute a principal source of public revenue and national wealth. Crude oil and natural gas, in particular, contribute significantly to the economies of resource-rich African countries (see Chapter 2).

Under the right circumstances, an "oil bonanza" or natural resource boom can be an important catalyst for growth and development. Unfortunately, in many African countries natural resource booms have set off dynamic growth processes only to a limited extent. The failure of natural resource wealth to lead to the expected economic growth and development has been attributed to several factors, including[1]

[1] See the *African Development Report 2007* (AfDB, 2007*a*) for a detailed literature review and analysis.

- "Dutch disease"—the syndrome of rising real exchange rates and wages driving out pre-existing export- and import-competing industries;
- rent seeking by elites and others who could otherwise put their energies into profit-making activities;
- price volatility and the "asymmetry of adjustment" (it is easier to ramp up public expenditure than to wind it down again);
- inflexibility in labor, product, and asset markets; and
- tensions between oil-producing and non-oil-producing regions within countries.

3.1.1 *What Is the Resource Curse?*

Essentially, the resource curse refers to the inverse association between development and natural resource abundance. It describes a situation whereby an export-oriented natural resources sector in a country generates large revenues for government but leads paradoxically to economic stagnation and political instability (Overseas Development Institute (ODI), 2006). It is commonly used to describe the negative development outcomes associated with non-renewable extractive resources (petroleum and other minerals). It has often been asserted that petroleum, in particular, brings trouble—waste, corruption, consumption, debt overhang, deterioration, falling apart of public services, wars, and other forms of conflicts, among others. Thus, growth in natural resource-abundant countries tends to be slower than expected—considering their resource wealth—and, in many cases, is actually slower than in resource-scarce countries.

A common thread in explaining the resource curse—along with the other broad explanations provided above—is the central role of government behavior. The key issue here is how governments administer resource wealth and how they use natural resource revenues.

3.2 Wealth Creation

3.2.1 *Oil and Gas Resources and Wealth Creation*

Three key factors underpin the facilitation and process of oil exploration and production: crude oil prices, technology, and fiscal regime. The legal framework that embodies the fiscal regime under which oil companies operate will—together with geological, geographical, and political factors—make a country or region more or less attractive for investors. The competitiveness of the regime is an essential determinant of the amount of upstream investment in exploration, discovery, and production by multinational oil companies. African countries are no exception to this conventional wisdom.

The rent from oil exploitation is very large. The sharing of this substantial rent between the governments of oil-producing and oil-consuming countries and oil companies has historically been contentious. The governments of both producing and consuming countries have tried to extract a sizable proportion of this rent. This was a key issue in OPEC countries' struggle for ascendancy in the world oil market in the 1970s (see Chapter 2).

A variety of upstream arrangements exist in many oil-producing countries. However, there are three major types of upstream arrangements: joint ventures (JV), production sharing arrangements (PSA), and service contracts (SC). Joint ventures are typically partnerships between state oil companies (SOCs) and multinational oil companies (MNOCs). In a JV type of petroleum agreement, the participants jointly share the risks of exploration and production in proportion to their equity share. The oil produced is also shared based on relative equity shares. Typically, the SOC has the majority share in the joint venture. A major problem with this arrangement, from the perspective of MNOCs, is the delay that often accompanies state oil companies' contribution to the joint venture exploration and production costs.

In a production sharing arrangement (PSA), the MNOC typically finances all exploration and development costs. If no oil is found in commercial quantity, the MNOC bears all the risks associated with exploration and development. If oil production commences and profitable extraction is achieved, the profit is shared between the company and the SOC/government according to an agreed formula. This option is much preferred by MNOCs because it gives them significant autonomy in operations and there is rapid recovery of upstream investment. Increasingly, oil-prospecting or oil-producing countries favor PSAs since this arrangement eliminates the need for the government to commit scarce funds upfront for the exploitation of its oil and gas resources.

The third major type of petroleum agreement is the service contract (SC) model. One sub-type involves a pure service contract, in which a company is contracted to undertake specific upstream activity—for example, drilling or seismic surveys—and is paid for this service. Another sub-type is "the risk service contract," under which the oil company undertakes all the exploration risks and is paid for its services at a fixed rate of return, if oil is found. Costs are reimbursed at an agreed markup over the investment in the oil discovered. However, if no oil is found, the company bears the full investment loss.

Structural changes in the world oil industry have brought significant changes to upstream petroleum agreements. Prior to the 1970s, when the MNOCs fully controlled the world oil industry, concession agreements dominated the world oil industry. However, in the 1970s when OPEC gained ascendancy over multinationals as the key player in the industry, the SC model assumed greater importance from the producing countries in exploitation of oil and gas in their domain. Total or partial nationalization became the dominant trend in oil-producing countries. New oil agreements emerged in response to the demands of producing countries and also in line with the prevailing economic nationalism of the era, implicit in the emergence of SCs. In recent times, similar "nationalization processes" have materialized in such countries as Venezuela and Russia.

The type of agreement selected in an oil-producing country should ideally depend on three important parameters: the size of reserves, the exploration and production costs, and the recovery factor. Risk service contracts seem to be dominant in countries with large reserves and low costs. Host countries with low costs and reserves tend to demand various levies to maximize the rent they can extract from oil. Production sharing arrangements dominate in countries with medium costs and large reserves. Joint venture royalty/tax systems are dominant in countries with low reserves and high costs. Furthermore, higher exploration and production costs and risks associated with limited reserves or deepwater offshore exploration often require more flexible fiscal terms for oil companies. The structure of royalty tax is "water depth dependent," as demonstrated in Table 3.1, which shows the cases of this relationship for Nigeria. Deep offshore oil exploration attracts no royalty payment.

More than one type of agreement can be available in a number of African countries. This largely reflects the diversity in costs and reserves in various oil fields in the countries—offshore and onshore— as well as the prevailing regime (policy) at the time the contractual arrangements are established, and so on. Again, Nigeria is an example of a country where the three major types of agreements operate simultaneously.

Table 3.1: Royalty Regime and Other Levies for Offshore Oil in Nigeria

Water depth in meters	Royalty (%)	Signature bonus ($ millions)
Below 100 meters	18.50	10
Up to 200 meters	16.67	10
Up to 500 meters	12.00	20
Up to 800 meters	8.00	25
Up to 1000 meters	4.00	20
Beyond 1000 meters	0.00	20

Data Source: NNPC (2008).

Table 3.2 provides information on the competitiveness of African oil-producing countries in relation to other oil-producing non-African countries. African countries fall into two categories: low cost and medium cost producers (for example, Algeria and Nigeria, respectively). Among Africa's major oil-producing countries, Angola has the highest exploration and production costs at $8, compared with $5.25 in Nigeria; $4.25 in Algeria; and $3 in Saudi Arabia. The range of upstream costs is from $3 at the low end to $14.5 at the upper end in the U.S. Gulf of Mexico.

3.2.2 *Management of Oil and Gas Resources*

This section presents a summary of the approaches adopted by African nations to manage, conserve, and enhance their oil and gas resources. It also highlights

Table 3.2: Petroleum Exploration and Production Costs (US$, year 2000)

Country	Exploration Cost	Production Cost	Total Cost
Low cost			
Saudi Arabia	1.50	1.50	3.00
Kuwait	1.75	1.80	3.55
Iraq	2.25	1.50	3.75
Iran	1.75	2.50	4.25
Venezuela	1.20	3.42	4.62
Algeria	2.15	2.50	4.65
Medium cost			
United Arab Emirates	3.00	1.80	4.80
Kazakhstan	3.50	1.30	4.80
Nigeria	3.00	2.25	5.25
Oman	3.75	2.50	6.35
Brazil	3.80	3.20	7.00
China	3.50	4.00	7.50
Russia	4.25	3.50	7.75
Angola	5.00	3.00	8.00
Indonesia	2.50	6.00	8.50
High cost			
U.S. Lowe-48 onshore	4.95	3.57	8.52
Canada Western	6.75	3.00	9.75
North Sea	7.50	3.00	10.50
Canada Eastern	8.00	3.80	11.80
U.S. Gulf of Mexico	11.00	3.50	14.50

Source: Al-Attar and Alomair (2005).

the benefits and wealth these countries can reap from utilizing these resources (AfDB, 2007).[2]

Overall, management of non-renewable resources involves

[2] A detailed analysis is presented in the *African Development Report* 2007 (AfDB, 2007*a*).

- ensuring the availability (exploration and extraction) of the resources;
- allocating resources to competing players—this may entail participation of local versus international players;
- creating an environment for resource industries to flourish;
- ensuring integrity in the management of revenues received from the extraction, mining, and processing of resources;
- developing policies to manage national ownership of non-renewable resources;
- limiting the environmental impact of resource exploitation;
- ensuring health and safety in the process of resource exploitation;
- converting resource use into sustainable economic development through linkages; and
- using resource rents for economic and social capital development and ensuring the overall creation of wealth and well-being in the country.

One of the key concerns is that African governments that are major producers of fossil fuels (and other minerals) do not receive sufficiently large rents or revenues from the production of these extractive products. This is attributable to a number of reasons, including contracts and regimes that are not designed to extract maximum rents; and mineral policies that are designed primarily to promote and attract investments and have not evolved with changing global dynamics and national interests.

Mineral resource rent is the value of mineral resources minus all the necessary production costs. A tax system can be designed to capture rents using conditional payments—payments that are closely linked to actually realized rents. That is, one pays if one meets the criteria to pay. The fiscal system must offer incentives to explore and invest while securing a fair share of revenues for public use. The mineral tax system cannot move too far out of line with those countries of similar "prospectivity" (degree of assurance to find and extract at economically competitive and viable conditions). It can be structured to reduce investment risks and secure higher government revenues. A mineral resource tax system can differ from the general tax system and still remain neutral. The system can use production sharing, taxes and royalties, or degrees of state-ownership, each with equivalent fiscal effect. Investors will be interested in the overall impact of the tax regime under a range of assumptions about output costs and prices. This impact has two aspects: The tax burden and the tax structure. The tax burden—specifically the average tax burden—is the share the government takes in taxes, duties, and royalties over the life of a resource field (that is, the period when resources are extracted from an oil or gas field). But the tax structure is the way in which the tax burden is imposed at different points in the life of the resource field. In appraising a mineral resource project, large companies will first examine the intrinsic economics of the project under the given fiscal regime. This usually involves estimation of an expected rate of return in discounted cash flow terms, in constant prices, in an all-equity case. This return will have to exceed the corporate threshold adjusted for special project and political risks. The average tax burden is vital to this assessment, but so is the timing of the major part of the tax burden and thus the tax structure. For the latter, when a given tax burden is imposed, the investor's expected rate of return must be high enough to encourage the investor to invest. Expectations of fiscal stability will reduce investors' perceptions of risk and increase the rents that the state can secure (ESMAP, 2004).

Governments need to devise and implement appropriate and modern tax regimes for mineral resource extraction. One of the challenges is that governments rarely believe that companies pay too much tax; companies rarely believe they pay too little tax; and citizens rarely believe that they actually see the full benefits from the taxes that are paid.

One of the main forms of government income from mineral resource exploitation is **royalties**, a taxation form that has created controversy across the globe, unlike any other form of mineral resource tax (Otto and Cordes, 2004). Other common tax forms that cumulatively comprise "**the government take**" include state tax, revenue tax, corporate tax, income tax, production sharing, state equity (part ownership), and various special mineral resource taxes (ESMAP, 2004).

An analysis of fiscal regimes (policies, legal framework, fiscal system, revenue collection, and so on) for fossil fuels in Africa reveals that these regimes are by no means uniform. A multitude of royalties, taxes, resource rents, incentives, state equity levels, and so on, have been developed to foster interest in exploration and investments, on the one hand, and to capture some of the benefits for the state and the public, on the other hand. Table 3.3 outlines some of the key characteristics of fiscal regimes in Africa. As clearly demonstrated in the table, the "taxation levels" and principles applied are as heterogeneous as the landscape and people in Africa.

3.2.3 *Sustainable Resource Development*

Sustainable development of non-renewable resources, including oil and gas, encompasses all the policies, principles, and practices that support the utilization of resources in a manner that does not prevent future generations from accessing the resource(s) or its benefits (UNECA and AfDB, 2007). A key purpose is to ensure that mineral-hosting nations in Africa benefit from their mineral resource endowments in the short and long terms, for example, by using the revenues accrued from mineral resource development for socioeconomic development

Table 3.3: Key Characteristics of Fiscal Petroleum Regimes in African Countries

Country	Royalties %	Production sharing %	Income tax rate %	Resource rent tax	Investment incentives (1)	State equity (2) %
Angola	. . .	15–80	50	None	Yes (E)	25
Cameroon	Negotiable	None	48.65	None	Yes (O)	50 (C)
Chad	12.5	None	50	None	None	10
Gabon	10–20	65–85	Gov. Share	None	Yes (E)	15 (C)
Mozambique	8	10–50	40	None	Yes (E)	None
Niger	12.5	None	45	None	Yes (E)	. . .
Nigeria	0–20	20–65	50, 85	None	Yes (E, Cr)	Variable
Sudan	None	60–80	None	None	. . .	None
Algeria	10–20	60–88	Gov. Share	None	None	30 (C)
Egypt	10	70–87	Gov. Share	None	Yes (I)	None
Libya	16.67	5–90	None	None	. . .	. . .
Tunisia	2–15	None	50–75	Yes	Yes (E, U, I)	Negotiable

Notes: (1) Investment incentives: tax holiday (H), accelerated depreciation (A), tax credit (Cr), current expensing of exploration and/or development cost (E), duty exemption for imports of equipment and capital goods (I), unlimited loss carry-forward (U) and other (O).
(2) The maximum equity share that the state can elect to take, often on a carried basis (C).
Source: Adapted from ESMAP (2004).

programs, creation of manufacturing industries, and other initiatives (Rogers, 2007).

The following specific principles, which are noticeably absent in the majority of African countries, can improve and enhance extraction to ensure more sustainable development (Otto and Cordes, 2004; Rogers, 2007):

- preserving strategic minerals of importance for future development (and generations);
- enforcing production quotas or caps;
- limiting the number of exploration licenses used, the areas available for exploration, or the number of extraction sites;
- ensuring longer production life by limiting annual capacity;
- establishing a profits trust framework; and
- instituting incentives to promote potential alternatives.

The majority of countries have taken important steps to formulate policies and legislation, and to incorporate fiscal terms into their strategies (AfDB, 2007*a*). However, more coherent principles, structures, and, above all, due diligence in enforcement would considerably increase benefits and sustainability for all countries. Some concrete issues related to sustainable development need to be improved. These include incorporating environmental aspects into the full extraction cycle and securing proper rehabilitation. Above all, although the countries have the means and measures in place to secure significant economic and social benefits from fossil fuels and mineral exploitation, a number of key questions remain:

- Are African countries benefiting enough from the resources at hand?
- Is the wealth created reaching the poor and the general population to a sufficient degree?

3.3 Resource Curse: The African Context and Experience

For oil and gas producers and exporters, higher oil prices are expected to be a blessing rather than a curse. Yet the evidence shows that in many of the net oil-exporting countries (in Africa and elsewhere), it has been a major source of economic, social, political, and environmental problems, rather than a treasure. The co-existence of significant oil wealth and large-scale poverty is certainly a "paradox of plenty," if not an outright resource curse in itself.

The essential issue is that to create and sustain long-term wealth—rather than a short-term oil boom—mineral resources have to be converted into other forms of capital (human, financial, and infrastructure) and more sustainable livelihood opportunities.

Drawing on an appropriate theoretical framework (see also AfDB, 2007*a*), this chapter examines the *African evidence* of the paradox of plenty and the "resource curse" with a view to exploring the following issues and questions:

- Is natural resource abundance—in particular, fossil fuels—in Africa a curse or a blessing?
- Has the management of natural resources really stunted the growth and development prospects of many resource-rich African economies?
- How does volatility in the export value of resources contribute to volatility in growth in GDP per capita and to long-term growth and development in general?
- What political and social factors allow some resource-abundant countries to utilize their natural resources to promote development and prevent others from doing so?
- How, or why, has the potential resource curse been avoided in some cases and how can it be overcome in the future?

In most of the analyses that follow, African countries are categorized as

1. Resource-rich (encompassing both oil and mineral exporters);
2. Oil-rich (oil and gas);
3. Mineral-rich (other minerals, such as metals); and
4. Resource-scarce.

Africa has 22 resource-rich countries, defined in this study as countries where fuel and mineral exports contribute over 20 percent to GDP. These countries account for slightly more than two-thirds of Africa's GDP and for half of its population (Table 3.4). Half of these countries are oil exporters (fossil fuels), while the other half are mineral exporters (metals and other minerals). In contrast, there are 31 resource-scarce countries, which account for 30 percent and 48 percent of regional GDP and population, respectively.

3.3.1 *High Export Dependence in Africa's Resource-Rich Countries*

Africa's resource-rich countries continue to be highly dependent on natural resource exports for both foreign exchange and revenues. For example, fuels accounted for 65 percent of the total increase in export values in African countries between 2000 and 2005. Since 1990, the share of fuels in the total exports of African oil-exporting countries has increased by about 12 percentage points, to almost 90 percent (IMF, 2006).

In the last five years, buoyant oil, gas, and mineral price increases have enabled resource-rich African countries to increase their natural resource exports and thus their revenues substantially. These increased revenues are a significant source of income for these countries, demonstrating the importance of natural resources in output growth and capacity to generate export revenues. For example, oil revenues account for more than half of all revenues in Angola, Congo, Equatorial Guinea, Gabon, and Nigeria. In fact, in U.S. dollar terms oil

Table 3.4: Macroeconomic Indicators by Resource Endowment Category, 2007 or Most Recent Data

	Population (million)		GDP Per Capita (current US$)	Real GDP Growth (annual %)	GDP (PPP)		FDI (Cur, US $m)	
	Pop in Million	% Share of Africa			Billion US$	% Share of Africa	Million US$	% Share of Africa
Oil-exporting countries	373	38.75	53,842	5.9	1,272	52.53	24,949	70.19
Mineral-exporting countries	263	27.26	20,045	4.3	660	27.25	2,005	5.64
Others	328	33.99	33,900	4.8	489	20.21	8,590	24.17
Africa	964	100.00	1,290	5.8	2,421	100.00	35,544	100.00

Source: Authors' calculations based on the AfDB Socio Economic Database.

revenues increased about 3.5 times between 2002 and 2006. In addition to the upsurge in revenue, production also expanded significantly, by 45 percent on average, especially in Angola, Chad, and Equatorial Guinea (IMF, 2006). Thus, oil-exporting countries, in particular, are highly revenue (fiscal) dependent on oil, implying the need for prudent fiscal discipline to minimize the adverse effects of the boom-bust cycle of oil prices.

The data shows that government expenditures have risen in recent years, but not at nearly the same rate as natural resource revenues. Before the current boom, non-oil deficits exceeded oil revenues in many resource-rich African countries (such as in Angola, Congo, and Nigeria); since then, the ratio of non-oil fiscal deficits to oil revenues has improved noticeably (Table 3.5). This reflects both the rapid rise in oil revenues and the narrowing of non-oil fiscal balances.

Table 3.5: Fiscal Balance, Investment Rates, and Terms of Trade Changes (in %)

	2000–2003			2004–2007		
	TOT	FD	INV	TOT	FD	INV
Oil-exporting countries	6.70	1.06	21.02	8.23	7.71	21.77
Mineral-exporting countries	−0.93	−4.14	16.33	4.11	−0.32	20.39
Other	1.77	−5.35	22.61	−1.94	−1.22	24.70
Africa	2.78	−1.83	19.87	5.84	2.43	21.85

FD: Fiscal Deficit GDP; INV: Domestic investment growth; TOT: Change in Terms of Trade

Source: Authors' calculations based on the AfDB Socio Economic Database.

The relatively cautious fiscal policies in many resource-rich African countries are helping these countries reduce their macroeconomic vulnerabilities. In other words, a good number of countries have used natural resource revenues to strengthen their external positions by reducing external debt (especially Gabon and Nigeria); accumulating external reserves (Angola, Congo, Equatorial Guinea, Gabon, and Nigeria); and reducing domestic and external arrears (Angola, Equatorial Guinea, Gabon, and Nigeria). Cameroon, Angola, and Congo have also improved their non-oil primary fiscal balances (IMF, 2007).

3.3.2 *Foreign Direct Investment in Resource-Rich Countries*

A major concern about foreign direct investment (FDI) to Africa is that the overwhelming majority of the inflows go into natural resource exploitation. Most of the flows to the top recipient countries—Angola, Algeria, Sudan, Nigeria, and Gabon—go to oil and gas projects.

3.3.3 *Growth Performance: Resource-Rich Countries*

Before the first oil shock in the 1970s, average oil-rich African countries enjoyed favorable macroeconomic conditions: robust economic growth, moderate inflation, manageable fiscal deficits and external debt, and external current account surpluses. The pro-cyclical policies they followed during the oil boom of the 1970s were intended to use the oil bonanza for economic and social development and to encourage economic diversification. Unfortunately, these objectives were not achieved since the actual results were economic imbalances that caused major distress when oil prices plunged in the 1980s and stayed low for over a decade (IMF, 2007).

Indeed, as Figure 3.1 shows, resource-rich Africa experienced a very disappointing two decades—1980–1999. However, a reversal has occurred in the last five to ten years, and current growth rates are in fact trending upwards. Cumulatively, resource-rich countries only experienced an average growth rate of 2.4 percent from 1981 to 2006, considerably lower than the 3.8 percent

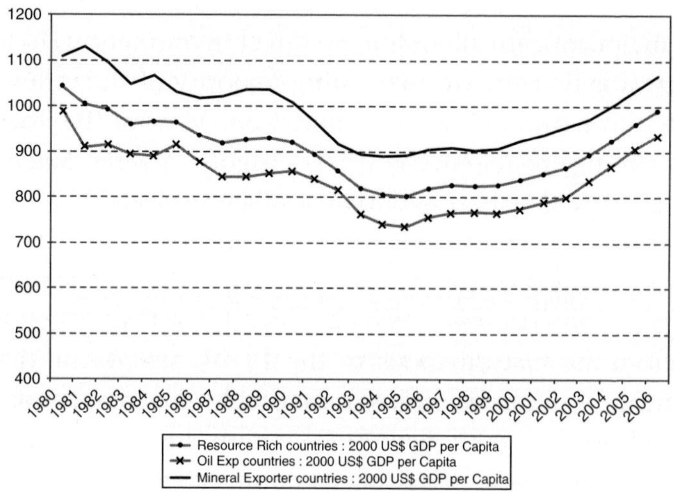

Figure 3.1: Natural Resource Abundance and Real GDP Per Capita

Source: AfDB Statistical Department, 2007; Computed from database.

average for resource-scarce countries. Resource-rich African countries are, nevertheless, richer than their resource-scarce peers (US$991 versus US$671 per capita, as Table 3.4 shows). The gap narrowed considerably during the 1980–2000 period but is widening again in conjunction with the recent resource boom (it has, all-in-all, narrowed from approximately 50 percent to 30 percent).

3.3.4 *Savings in Resource-Rich Countries*

One of the features of many countries that are endowed with abundant natural resources is that they save less than what is expected, considering the rents obtained from extracting and selling natural resources. Presumably, if the countries saved more, they would grow at a sustainable and faster rate. To gain a better understanding of sustainable development, it is useful to examine the concept of *genuine* saving.[3]

Genuine saving corresponds to an increase in the wealth of a nation. According to the so-called Hartwick rule, any depletion of natural resources or damage by stock pollutants must be compensated for by increases in non-human and/or human capital. This rule of zero genuine saving can be seen as a rule of thumb or motivated by max–min egalitarianism. It requires that resource-rich

[3] Genuine saving is defined as public and private saving at home and abroad, net of depreciation, *plus* current spending on education to capture changes in intangible human capital, *minus* depletion of natural exhaustible and renewable resources, *minus* damage of stock pollutants (CO_2 and particulate matter).

countries adopt a strategy for transforming their natural resource wealth into other forms of productive capital (World Bank, 2006*c*; and Ploeg, 2007).

Resource-rich countries in Africa therefore need credible and transparent rules for sustainable consumption and investment to ensure that exhaustible natural resources are gradually transformed into productive assets at home or abroad. Furthermore, countries with high population growth rates need *positive* rather than *zero* genuine saving rates to maintain constant consumption per head. They thus need to save more than their exhaustible resource rents—but only rarely manage to do so.

Analysis suggests that resource-rich countries with negative genuine saving, such as Nigeria, would experience increases in productive capital by a factor of five or four if the Hartwick rule were applied. Effectively, for countries with negative genuine saving, the erosion of their natural resource wealth exceeds their accumulation of other assets. They squander their natural resources at the expense of future generations without investing in other forms of intangible or productive wealth. This is an unfortunate feature of several resource-rich African economies (Figure 3.2).

3.3.5 *Human Capital Development and Income Distribution*

One of the dilemmas of natural resource abundance is that it *may,* pervasively, cause a country to neglect

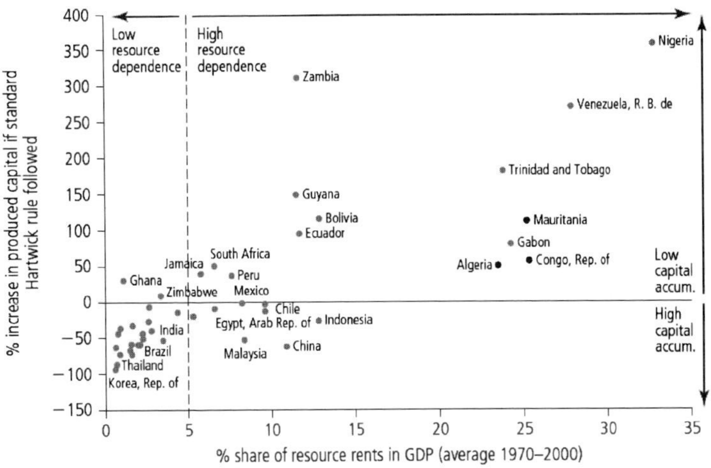

Figure 3.2: Counterfactual Exercise—Imposing the Hartwick Rule

Source: World Bank (2006*c*), "Where is the Wealth of Nations?"

human capital development—the same basic causes and effects outlined above in reference to negative genuine saving. High levels of natural resource revenues can thus divert attention from diversification and wealth creation, including from institutional and human development (World Bank, 2006*c*; and Ploeg, 2007). The logical expression of such a potential correlation between resource abundance and neglect of human capital development would, in the medium to long term, be reflected in a low basic human development status. The United Nations Human Development Index (HDI), a comparative measure of life expectancy, literacy, education, and standard of living in countries worldwide, provides a standard means of measuring human well-being and country

Table 3.6: Resource Abundance and Social Performance

	Human Development Index (HDI) (Scale 0–1; Niger lowest with 0.31; Seychelles highest with 0.84)	**Income Inequality (GINI Coefficient Index; scale 0–100)***
1- Resource-rich countries	0.51	31.1
Oil-exporting countries	0.55	15.3
Mineral-exporting countries	0.46	46.8
2- Resource-scarce countries	0.51	26.8
5- Africa	0.51	45.9

* For the GINI Coefficient Index; 0 corresponds to perfect equality
Sources: UNDP (2006), Human Development Report 2006; World Bank (2006*d*).

development status. As reported by the UNDP in its 2006 Human Development Report, Africa dominates the low end of the HDI (29 of the 31 countries with a low human development status). Only the Island States of Seychelles and Mauritius qualify as having a high human development status. The remaining 22 countries, including all the North African Arab states, have a middle-level human development status. It is also worth noting that oil-rich Norway has the highest HDI among all countries in the world (UNDP, 2006).

A deeper analysis of HDI data (Table 3.6) indicates that there is no difference between resource-rich and resource-scarce countries (0.51). However, oil-rich countries are doing considerably better in this aspect than primarily mineral-rich countries (0.55 compared with 0.46).

Another important aspect, frequently highlighted as problematic in resource-rich countries, is increased income inequality. Oil, gas, and mining industries are often characterized by their "enclave nature," with few forward and backward linkages into the economy. During exploration and production, such industries employ only a relatively small number of highly-skilled, well-paid workers and generally import the majority of inputs. Furthermore, there is a considerable risk that public expenditure during a resource boom may exacerbate inequality, for example, concentrating expenditure in the formal sector in towns and cities, skewing distribution (not benefiting rural households), and prioritizing the interests of the elites and wealthier classes. Because of these tendencies, society tends to identify the production and export of natural resources with the interests of the rich (Overseas Development Institute, 2006). As Table 3.6 shows, income inequality is noticeably higher in resource-rich African countries (Gini Coefficient of 31.1) than in resource-scarce countries (Gini Coefficient of 26.8). Nevertheless, it is again worth noting that income inequality is more pronounced in mineral-rich countries than in oil-exporting countries.

3.4 Policies and Governance

In most countries and legal regimes, oil, gas, and minerals are vested to the state. Revenues in the first instance accrue to the government, inviting government action, in one way or another, to spend some of the accrued revenues. There is a recurrent debate on how or why this very often results in policy failures and poor governance. Several strands of arguments are presented below (see also the African Development Report—ADR, 2007).

3.4.1 *Poor Decision-Making*

The first strand argues that large windfall revenues lead to poor decision-making by governments. This is attributable to several factors, including the following (Ploeg, 2007; Auty, 2001; Auty, 2004; Stevens, 2003):

- Resource booms raise expectations and increase appetite for spending. The promise of natural resource wealth dramatically expands the horizons of governments in natural resource-exporting countries. A boom mentality not only affects the way governments behave—creating grandiose plans and ideas; it also shapes how people respond. Work ethics may be undermined, resulting in decline in productivity.
- The development of oil, gas, or minerals raises expectations among the population. This pressures governments to "do something," thus encouraging speedy responses. This often leads to quick, inappropriate, and poorly coordinated decisions.

- Having more money to "play with" tends to weaken prudence and normal procedures of "due diligence." In particular, governments may decide on capital spending without due thought to recurrent spending implications.
- Governments often dramatically increase public spending based on unrealistic revenue projections. In resource-dependent countries, windfalls increase both public spending and the appetite for transfers by a factor that is more than proportionate to the size of the boom itself. This means that spending quickly surpasses revenues. Nonetheless, various interest groups continue to demand even larger shares of national income when natural resource revenues go into a downtrend.

3.4.2 *Corruption and Rent Seeking*

It is often argued that natural resource booms often decrease the quality of public spending and encourage rent seeking (Ploeg, 2007; Auty, 2001; Auty, 2004; Stevens, 2003). Centralization, and, hence, concentration of fiscal resources from resource booms, fosters excessive and imprudent investment. It also leads to mismanagement and misallocation of resources and, in the most severe cases, corruption.

The key issue is that natural resource revenues tend to replace more stable and sustainable revenue streams, exacerbating problems of transparency and accountability. With sizeable resource revenues, the reliance on non-resource taxes and other government incomes decreases. This tends to free natural resource-exporting governments from the types of citizen demands for fiscal transparency and accountability that arise when people pay taxes directly to the government. Thus, natural resource export earnings actually sever important links between the people and their governments, links related to popular interests and control mechanisms.

The larger the public purse, the less noticeable the leakage to interest groups. Rent seeking is greater in resource-rich countries because wealth is concentrated in the public sector (or possibly in a small number of companies). Therefore, the bulk of the rents created in these economies are channeled by bureaucrats, the majority of whom are members of the politically dominant groups. Such rent-seeking behavior produces undesirable results for the economy. First, rent-seeking behavior imposes significant losses on the economy. Second, it distracts attention away from long-term development goals to maximizing rent creation and capture. Third, rent seeking creates extremely powerful lobby groups that are able to block needed economic reforms. Fourth, societies face severe impediments to innovation as a result of the behavior of special interest groups. Fifth, rent seeking makes it more difficult for governments to adjust spending when faced with revenue fluctuations. Finally, rent seeking is tantamount to the creation of monopoly power in an economy and the social costs of such monopolization are higher if the costs to maintain that monopoly are added (Stevens, 2003).

Governance indicators, such as government effectiveness, voice and accountability, political instability and violence, the rule of law, regulatory quality, and control of corruption are markedly weaker in oil-rich African countries (see Table 3.7). The data furthermore indicates that this problem is by far most common in relation to oil exploration and revenue, as the performance of oil-exporting countries is clearly worse than that of countries with other minerals.

The literature emphasizes the role of governments in the misallocation of resource revenues (Stevens, 2003). Resource booms have adverse effects because they provide incentives for politicians to engage in inefficient redistribution of revenues and income in return for political support. However, it is important to note that the status of existing institutions (before the resource boom) is crucial, as it determines the extent to which politicians can respond to these perverse incentives. Regardless of the starting point, pressure from the public to raise public spending is likely to be significant—leading to inefficient redistribution in the form of public employment provisions, subsidies to farmers, labor market regulations, and protection of domestic industries from international competition.

3.4.3 *State Predation*

The attainment of political independence did not transform the structure of a good number of African states, which remained forceful and authoritarian. Thus,

Table 3.7: Natural Resource Abundance and Governance Indicators

	Voice and accountability 2006	Political stability 2006	Government effectiveness 2006	Regulatory quality 2006	Rule of law 2006	Control of corruption 2006
1-Resource-rich countries	−0.8	−0.7	−0.8	−0.7	−0.9	−0.8
Oil-exporting countries	−1.3	−1.0	−1.0	−1.0	−1.1	−1.0
Mineral-exporting countries	−0.4	−0.3	−0.5	−0.5	−0.6	−0.5
2-Resource-scarce countries	−0.5	−0.4	−0.7	−0.7	−0.6	−0.5

Data Source: Kaufmann, Kraay and Mastruzzi (2007).

instead of transforming the state and making it relevant to the needs and aspirations of the populations, some emerging post-colonial leaders were content with using the enormous authoritarian structures of the state to appropriate economic gains for private ends. In response to the forces of globalization—epitomized by the end of the Cold War, growing pressures for both economic and political liberalization, as well as increasing internal resistance and demands for democratization—many state regimes have resorted to repression and predation. A predatory state is characterized by the concentration of power at the top and the personalization of networks for delegation of this power, which is enforced by ruthless repression. In this context, economic inducements for government officials and generalized corruption are the government way of life (Castells, 2000).

Predatory rule has two major consequences on natural resource wealth and revenue management in most fragile African states. First, access to state power is equivalent to access to wealth and to the sources of future wealth. Second, political support is built around clientele networks, which link power-holders with segments of the population. The concern of the various elites, ultimately connected to the top of state power, is how to gain support and consolidate clienteles while maximizing the amount of resources needed to obtain this support. These networks are formed along ethnic, regional, territorial, religious, and economic lines (Castells, 2000).

Predatory states use different violent and non-violent strategies to manage mineral and oil resources and to appropriate the proceeds accruing from their exploitation and sale. Since minerals are extracted in enclave production centers, sometimes located offshore, the common strategy is to negotiate royalties and other agreements directly with foreign companies. These deals are often shrouded in mystery, making it difficult, if not impossible, to track how much money is generated or how these revenues are spent. According to oil industry experts, OPEC countries on average retain some 75 percent of their oil revenues for the state budget, allowing for operating expenses. In the case of African oil producers, this proportion, even in the best-case scenario, falls in the range of 55 percent to 70 percent. The difference represents supplementary profits shared by the oil companies and African elites (Hibou, 1999).

3.4.4 *Sociocultural and Political Impacts*

As outlined above, countries that have abundant point-source natural resources (such as oil) tend to have less prudent policies and poor governance. They also tend to have weaker institutional capacities (Stevens, 2003). The following paragraphs discuss some of the reasons for this relationship and how it can be avoided.

Resource rents are an invitation to non-productive lobbying and rent seeking. This problem occurs mostly in countries with "grabber-friendly" institutions, while countries with "producer-friendly" institutions generally do not suffer from the curse. In other words, countries that avoided the resource curse in previous resource booms did so because they had transparent and sound institutions and because they adopted specific policies—including institutional capacity building—to minimize the impact of, and damage caused by, the resource windfalls.

Countries in which governance (hence transparency) is initially poor are thought to face a substantial risk of turning resource windfalls into catastrophe. Indeed, there is evidence that governance is likely to deteriorate further in a resource boom, even from a low-start status, because of the windfalls. Governance and effective public spending are thus critical for both living standards and private activity and, since the public sector is a large part of the economy, its own productivity growth is a key component of overall growth. This in turn requires that governments aspire to overall national goals, thus becoming accountable to citizens, regardless of their own interests and aspirations.

Poor economic performance during previous natural resource booms underscores the importance of sound macroeconomic policies and strong institutions. The large public investment projects of the 1970s and 1980s, when governance and institutions were extremely weak in most of Africa, were often undertaken with little scrutiny and accountability. The return on public investment was correspondingly low. Meanwhile, poor macroeconomic management of natural resource price cycles in several African countries resulted in large exchange rate appreciation, erosion of the competitiveness of non-oil sectors, and high inflation. Given that many African countries leveraged their natural resource wealth to access credit from foreign suppliers and governments, the early 1990s witnessed a sharp rise in external debt, well above 100 percent of GDP, in most cases, resulting in unsustainable external debt levels. These macroeconomic imbalances have eventually called for very painful policy adjustments, such as sharp fiscal contraction, trade liberalization, exchange rate adjustment, and debt rescheduling (Collier and Goderis, 2007).

In addition, with weak institutions and legal systems, there is a higher return on rent seeking, and a higher occurrence of crime, corruption, unfair company takeovers, and other shady dealings. A resource bonanza thus elicits more rent seekers and reduces the number of productive entrepreneurs. In the long run, profits fall, resulting in an economy that is worse off. Thus, if institutions are weak and conditions are not favorable, dependency on oil and on other natural resources effectively hinders democracy and the quality of governance (Ross, 1999; Ploeg, 2007).

Box 3.1: GOVERNANCE AND TRANSPARENCY IN NATURAL RESOURCE EXPLOITATION AND MANAGEMENT

Governance remains the overarching and most critical challenge for natural resource exploitation and management. Although African governments bear prime responsibility for managing natural-resource wealth in a fair, transparent, and accountable way, they are only one part of an intricate web of interests and relationships, which include regional actors, foreign governments, and multinational extractive companies. The main governance-related challenges facing resource-rich countries can be summarized as follows:

Transparency

Transparency is the key issue in establishing accountable governance structures and fighting corruption. However, this has to start with the concession contract itself, as well as with revenues accruing from the sale of the resources:

- Corruption in the allocation of resource concessions undermines governance in resource-rich countries and also entails a poor deal for their citizens. There is overwhelming evidence that concession allocation is often obscure and involves a lot of corruption.
- Concession contracts often contain confidentiality clauses and are therefore not open to public scrutiny. Without knowing the details of the deals signed by their government, the citizens of a given country have no way of holding their politicians accountable.
- Transparency is equally important for the revenue flows of natural-resource rents between extractive industry companies and host governments. If the companies publish what they pay and the governments publish what they earn, the revenue flows can be traced and governments can be held accountable for sustainable management of these revenues and fair distribution of the wealth.

Data Sources: Heinrich Boll Foundation (2007); Alley et al. (2007).

3.4.5 *Resource Governance Policy Initiatives*

The last decade has seen a growing recognition that improved transparency and accountability for the huge revenues generated by oil, gas, and mineral industries is vital to avoiding the "resource curse" and extending the benefits of natural resource abundance to poverty reduction. Consequently, several international policy initiatives, mechanisms, and standards have been launched to address these dilemmas, improve governance, and reduce the observed environmental and socioeconomic impacts of extractive industry activities. These include (see also AfDB, 2007*a*)

- The Extractive Industries Transparency Initiative (EITI)
- The international Publish What You Pay (PWYP) campaign
- The International Council on Mining and Metals (ICMM)
- The Kimberley Process Certification Scheme (KPCS)
- The Equator Principles (social and environmental standards)
- The African Peer Review Mechanism (APRM)

The most important of these—from an African perspective—are discussed below.

3.4.5.1 THE EXTRACTIVE INDUSTRIES TRANSPARENCY INITIATIVE

This initiative was launched by the UK government in 2002 to address the general failure to transform resource wealth into sustainable development (the "resource curse" or "paradox of plenty") and the associated governance problems in the extractive industries sector. The EITI aims to intervene in the middle of the value chain—collection of taxes and royalties stage—but neither upstream nor downstream. The EITI has grown into a worldwide initiative. More than 20 countries have committed to its principles and criteria, the majority of them African countries (see also www.eitransparency.org).

Assessment of African Participation: A considerable number of African countries have endorsed the EITI and are applying its principles to various extents. It is worth noting that the initiative is very much focused on the oil and gas sectors. Its scope and mandate can be expanded, revenue transparency strengthened, and the initiative extended to upstream and downstream issues and to environmental stewardship (Jourdan, 2006).

The Nigeria Extractive Industries Transparency Initiative (NEITI), operating the EITI "plus" model, is the most advanced and comprehensive of all the initiatives of participating countries. The NEITI was launched in 2004 and yielded the first-ever comprehensive financial, physical, and process audits of Nigeria's oil and gas sectors for the 1999–2004 period. Cameroon and Mauritania have appointed stakeholder committees at the highest levels of government, drawn up action plans, and offered workshops and training for civil society. These two countries have also issued two reports (www.eitransparency.org).

In summary, and looking ahead, the EITI has recorded some significant achievements, although it is voluntary in nature. It is increasingly being recognized as a *partial* solution to the problem of corruption in energy-rich developing countries. However, the EITI faces a number of challenges that need to be addressed (EITI, 2007; AfDB, 2007*b*):

1. Not all countries that have adhered to the EITI have started implementing it in its full extent. Many countries thus only show rhetorical commitment and only actually implement the initiative to a limited extent.

2. Civil society organizations play an important role in the EITI and in its implementation at the national level, where such a multi-stakeholder initiative is often difficult to realize. However, in a number of countries, records have emerged of civil society representatives being harassed.
3. The EITI does not address the core problems of corruption, mismanagement, and accountability comprehensively.

The success of the EITI as a concept is increasingly threatened by the lack of clarity about what it means in practice. These problems and issues must be addressed, and EITI is only a first step in the right direction. One of the undeniable effects of the EITI process is that it has raised international awareness that transparency in oil, gas, and mining revenues is vital to preventing corruption in countries that depend on resource revenues and to ensuring that these revenues are used to promote growth and development. The EITI has brought together companies, investors, governments, civil society groups, and international institutions to promote this shared vision.

3.4.5.2 THE INTERNATIONAL PUBLISH WHAT YOU PAY CAMPAIGN

The Publish What You Pay (PWYP) initiative is a coalition of over 300 global civil society organizations from more than 50 countries. It aims to promote full transparency in the payment, receipt, and management of revenues paid to resource-rich developing country governments by the oil, gas, and mining industries. PWYP campaigns are focused on achieving mandatory disclosure of payments made by oil, gas, and mining companies to governments and other public agencies. This is commonly accepted as a necessary first step towards a more accountable system for the management of natural resource revenues. Such disclosures will not only allow members of civil society to draw a comparison and thus hold their governments accountable for the management of revenues; it will also strengthen the social standing of companies by demonstrating their positive contribution to society.

Assessment of African Participation: The response of African countries to the PWYP initiative has been positive, with a considerable number of NGOs pledging their commitment. To date, NGOs from 23 African countries have joined the coalition. Nigeria has been a strong supporter of the initiative and is home to 47 PWYP coalition-member NGOs (PWYP, 2007).

The debate on PWYP (a voluntary principle) raises the same issues highlighted above in relation to the EITI initiative.

3.4.5.3 EQUATOR BANK PRINCIPLES

Ideally, banks and financial institutions should be made accountable for oil- and mineral-backed loans and disbursements, particularly when they undermine attempts by the international community and international financial institutions to control the flow of money that involves corrupt natural-resource deals. The two most well-known global banking initiatives that target this issue are the "Equator Principles," which set social and environmental standards for project finance deals, and the "Wolfsberg Group," which has developed a set of anti-money laundering principles. However, both of these initiatives are voluntary.

The Equator Principles are essentially a voluntary financial industry benchmark for determining, assessing, and managing social and environmental risks in project financing. Institutions that have adhered to the Principles (known as "Equator Principles Financial Institutions" or "EPFIs") have consequently adopted these values and principles. The initiative is currently supported by some 50 banks from 16 countries. In their current form, the Principles are based on the environmental and social safeguard policies of the International Finance Corporation (IFC). As such, they do not provide any specific guidance for extractive sector projects—a considerable limitation in itself. Nevertheless, the EPFIs have committed themselves to not providing loans to projects where the borrower is unable to address and comply with the general principles and overall stated policies and procedures (www.equatorprinciples.com).

A distinct disadvantage of the Equator Principles is that they are driven by the Western world and do not, as such, focus on developmental impacts in the developing world (Jourdan, 2007). It would thus be beneficial for financial institutions to look at developmental aspects as well. Given the important role that banks play in financing private sector projects around the world, it is critical that signatories to the Equator Principles join in the global push for transparency in extractive sectors. What is further needed, to boost the effect of this and similar initiatives, is mandatory transparency in the financing of resource projects. This ideally implies putting an end to resource-backed loans for governments, investors, and others who refuse to manage resource revenues in a transparent manner. This would also call for an amendment of money laundering regulations, recognizing that resource deals and resource-backed loans constitute a significant "yellow flag" for potential money laundering (Jourdan, 2007).

3.4.5.4 AFRICAN PEER REVIEW MECHANISM (APRM)

The African Union's New Partnership for Africa's Development (NEPAD) identifies good governance as a basic requirement for peace, security, and sustainable growth and development. One of its "immediate desired outcomes" is that "Africa adopts and implements principles of democracy and good political, economic, and corporate governance, and the protection of human rights becomes entrenched in every African country." For this purpose, NEPAD set up the African Peer Review Mechanism (APRM), an innovative tool aimed at peer review of governance benchmarks and design of action plans for improvement.

Participation in the system is voluntary and a panel appointed by the APRM Secretariat oversees implementation throughout Africa. The process consists of a number of stages (website: www.aprm.org.za). The ultimate stated goal of the APRM is to encourage African countries to plan a way forward on governance issues and implement relevant plans in this direction. The APRM process is designed to help participating countries develop and promote the adoption of laws, policies, and practices that lead to political stability, high rates of economic growth, sustainable development, and continental economic integration.

Assessment of African Participation: At present, 27 countries have committed to the APRM and 13 of them have had reviews launched. To date, Ghana, Rwanda, and Kenya are the three countries that have completed the entire process. The broad inclusiveness of the process has demonstrated the presence in countries of a strengthened culture of political dialog and empowerment. However, implementation has posed some challenges, for instance, the establishment of an appropriate national structure, the financing of the process, and the organization of a participatory and all-inclusive self-assessment system (APRM, 2007; AfDB, 2007*c*).

Voluntary participation in the APRM assessment has resulted in very high expectations, and it is imperative for member countries and stakeholders to now see the "dividends" from the APRM in terms of enhanced governance and improved living standards. The APRM is a cornerstone of NEPAD, the first African-initiated and -led instrument to take full ownership of the continent's future by addressing the key governance challenges that constitute major constraints.

3.5 Summary and Conclusion

The conclusion from the analysis in this chapter is that the true potential benefits of having significant natural resource wealth have *not* been fully exploited by oil-rich African countries. Overall, the performance of resource-rich countries in Africa has been disappointing, especially from 1980 to 2000.

However, despite the challenges and issues involved, a natural resource boom can, under the right circumstances, be an important catalyst for growth and development. The often referred to "natural resource curse" can be avoided with the right knowledge, institutions, and policies. Several African countries have demonstrated this, and there is reason for cautious optimism that more countries have learned hard lessons from past resource booms, and, in future, will pursue strategies and policies that will allow them to fully reap the benefits of natural resource wealth.

4

Impact of High Oil Prices on African Economies

This chapter reviews evidence of the economic and socio-environmental effects of high and rising oil prices on African countries. In the past, significant increases in the price of oil have led to worldwide economic recessions, such as the 1973 and 1979 energy crises. In many European countries, which have high taxes on fuels, such price shocks could potentially be mitigated by reducing the taxes as fuel costs rise. In many African countries, on the contrary, this option is not feasible, as there is limited taxation on oil products in the first place. In some cases, the oil prices are subsidized. African governments thus face a considerable dilemma: how much of the increase in oil product prices they should "pass through" to consumers. The negative effect of rising oil prices is thus potentially large for net oil-importing countries. In principle, the net effect should be positive for net oil-exporting countries, as positive effects offset negative effects. This chapter provides a comprehensive analysis of these issues.

4.1 Impact of Oil Prices on Africa

This section reviews and discusses the impact of high oil prices on African economies. The importance of oil products for African nations and the marked tendency for significant oil price volatility were presented in Chapter 2. Only a brief account is thus provided here. This section focuses primarily on estimating and outlining the effect of these developments on Africa's socioeconomic development.[1]

A barrel of crude oil was trading between US$18 and US$23 in the 1990s; it crossed the US$40 mark in 2004; and rose to about US$60 from 2005. During the summer of 2007, the price of one barrel of crude oil jumped above US$70 and even crossed the US$147 mark in July 2008. Although oil prices are still lower than in the late 1970s and early 1980s in real terms (inflation adjusted), the recent upsurge can have dramatic consequences on oil-importing countries. The impact of high oil prices is likely to be even more severe in countries that are overly dependent on oil and/or have limited access to international capital markets. In principle, net oil-exporting countries stand to benefit from the influx

[1] The review in this chapter is based primarily on AfDB Research Department documents (2007); ESMAP (2006); and several (unpublished) Research Papers by the AfDB Research Department.

of foreign revenue, which they can harness to accelerate the implementation of poverty reduction strategies and the attainment of the Millennium Development Goals (MDGs). However, these countries face challenges in managing the oil windfalls for the benefit of the whole population, as well as for future generations, and in cushioning their economies against any Dutch disease effects (see Chapter 3).

For many developing countries, the increase in oil prices over the last few years has made structural reform of the domestic petroleum pricing system a critical component of their macroeconomic policies. Although in some countries oil price increases may have been partly offset by exchange rate movements (notably the weakening of the U.S. dollar against the euro), it has also had major socioeconomic impacts. Many governments have been reluctant to pass on to consumers a rise in international oil prices because of the potential for social resistance to a policy that could hurt the poor. However, if they do not pass on the higher prices, their countries could experience a significant fiscal burden, which, in turn, could oblige governments to cut social spending.

Africa has 38 net oil-importing countries (see Chapters 2 and 3). As outlined above, in these countries in particular, high oil prices have an adverse impact on businesses, consumers, and the government budget, to name a few. As a result, these net oil-importing countries see their terms of trade deteriorate—jeopardizing their balance of payments positions and possibly leading to lower economic growth than in the absence of the oil shock.

4.1.1 *Literature Review*

A large number of studies have investigated the macroeconomic impact of oil price shocks, focusing in particular on the response of economic growth and consumer price inflation in oil-importing countries (see Barsky and Kilian, 2004; Hamilton, 2005; and Kpodar, 2006). A somewhat smaller body of literature has studied the impact of oil price shocks on external accounts (Bruno and Sachs, 1982; Ostry and Reinhart, 1992; and Gavin, 1990; and 1992), while yet other studies have concentrated on other relevant issues related to oil price shocks.[2]

There are fewer studies on the impact of high oil prices on African economies compared with other continents. However, specific studies have been conducted on the effects in Kenya, Nigeria, Mali, Mozambique, and

[2] These include the following: Krichene (2006) focused on monetary policy and the recent oil shock in the world crude oil markets; Hunt et al. (2001) studied the macroeconomic effects of high oil prices; Hamilton (2000) defined oil shocks and their effect on the economy; Daniel (2001) analyzed the possible solutions for high oil prices and market risks for governments; Hossain (2003) studied the taxation and pricing of petroleum products in developing countries; Kilian et al. (2007) looked at the relationship between oil price shocks and external balances; and Hunt (2005) studied the role of oil price shocks in the stagflation in the 1970s.

Ghana,[3] for example, while more general and continent-wide studies have gained attention in the last few years.[4]

4.1.2 *Impact of High Oil Prices on GDP*

There are two principal drivers of the demand for any product: growth in demand and the price of the product. These two drivers can in turn be divided into four factors: (i) The rate of growth of income; (ii) The income elasticity of demand for the product; (iii) The rate of increase of prices; and (iv) The price elasticity of demand for the product. Unfortunately, evidence on price and income elasticity in developing countries is sparse, partly because of lack of data, and tends to be based on data that is several years old (ESMAP, 2006).

Although the GDP of a country is a generic measure of economic performance, it gives a fairly good indication of the magnitude of the impact of oil prices on the economy of a given group of countries (see Table 4.1).

[3] Studies on the impact of high oil prices in African economies include works by the following: Semboja (1994) studied the effects of oil price changes in Kenya; Ayadi et al. (2000) and Ayadi (2005) studied the effects of oil production shocks in Nigeria; Kpodar (2006) looked at the way international oil prices are transferred to household expenditures in Mali; Nicholson et al. (2003) and Coady and Newhouse (2005) explored the distributional impact of an increase in the price of oil in Mozambique and Ghana, respectively; and Coady et al. (2006) studied the magnitude and distribution of fuel subsidies in developing countries, including in Ghana and Mali.

[4] This includes Bouakez and Vencatachellum (2007); and ESMAP (2006), which both address the impact of high oil prices in net oil-importing and net oil-exporting countries.

Table 4.1: African GDP Growth Rates, by Country Type, 1996–2005

Slow-growth economies GDP growth less than 4 percent a year (36.7 percent of population)		Diversified, sustained-growth economies GDP growth 4 percent a year or more (35.6 percent of population)		Oil exporters (27.7 percent of population)	
Country	**GDP growth (percent)**	**Country**	**GDP growth (percent)**	**Country**	**GDP growth (percent)**
Zambia	3.80	Mozambique	8.3	Equatorial Guinea	30.8
Guinea	3.70	Rwanda	7.6	Chad	9.0
Niger	3.50	São Tomé and Principe	7.1	Angola	8.5
Malawi	3.30	Botswana	6.7	Sudan	6.3
Mauritania	3.30	Uganda	6.1	Nigeria	4.3
Togo	3.30	Cape Verde	5.8	Congo, Rep.	3.4
Madagascar	3.20	Mali	5.8	Gabon	1.1
Lesotho	3.00	Tanzania	5.3		
Kenya	2.90	Ethiopia	5.2		
Eritrea	2.41	Sierra Leone	5.2		
Seychelles	2.30	Burkina Faso	5.0		
Comoros	2.13	Mauritius	4.8		
Central African Republic	0.85	Ghana	4.7		
Guinea-Bissau	0.47	Benin	4.6		
Burundi	0.43	Senegal	4.5		
Congo, Dem. Rep.	0.08	Cameroon	4.2		
Zimbabwe	−2.20	Gambia, The	4.2		
		Namibia	4.0		

Note: GDP growth rates are compound annual averages.
Source: World Bank Development Data Platform.

In Africa, 36.7 percent of the population lives in slow economic growth countries (GDP growing less than 4 percent a year); 35.6 percent in diversified and sustainable growth economies (GDP growing more than 4 percent a year); and only 27.7 percent in oil-exporting countries. This suggests that a 1 percent loss in GDP as a result of high oil prices will have a greater impact on slow growth economies and, consequently, on the majority of Africa's population.

Figures 4.1 to 4.5 show the GDP growth rate between 1999 and 2004 in selected countries by sub-region in Africa. In general, most countries show positive growth between the two periods analyzed (1990–1999 and 2000–2004). Although the growth pattern varies by period, it is clear that some countries recorded higher growth at the turn of the millennium (for example, the Democratic Republic of Congo, Rwanda, Comoros, Kenya, Tanzania, Algeria, Morocco, Tunisia, Sudan, Angola, Zambia, Mozambique, Botswana, Namibia, and South Africa). It is also worth noting that very few countries had outright negative economic growth (Central African Republic, Seychelles, and Zimbabwe).

Improved trade performance, particularly among oil exporters, is considered a major factor of economic growth in Africa in 2004 and subsequent years (AfDB, 2005). In 2004 oil prices rose 60 percent as a result of increased global demand, especially in China and Asia. According to the AfDB (2005), high oil prices are attributable to a rise in global demand rather than to supply-side factors. Nevertheless, supply-side factors such as

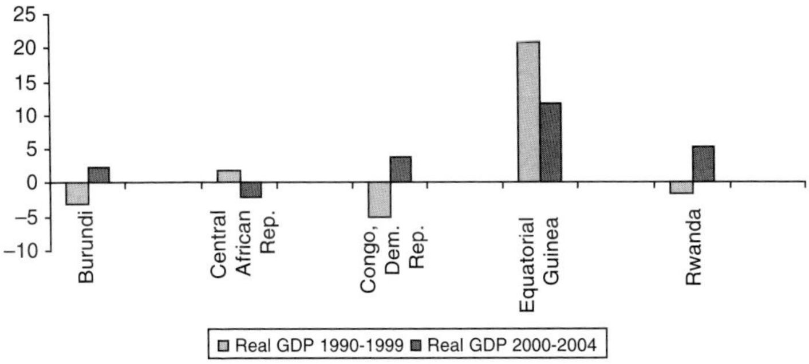

Figure 4.1: Average Annual Real GDP Growth in Central Africa

Source: IMF (2006); and AfDB Statistics Department Database (2007).

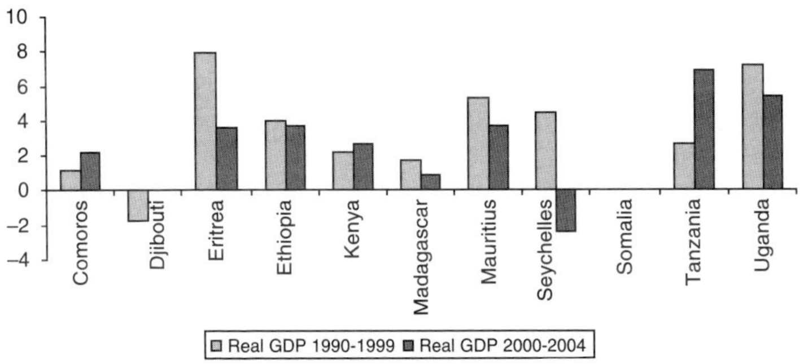

Figure 4.2: Average Annual Real GDP Growth in East Africa

Source: IMF (2006); and AfDB Statistics Department Database (2007).

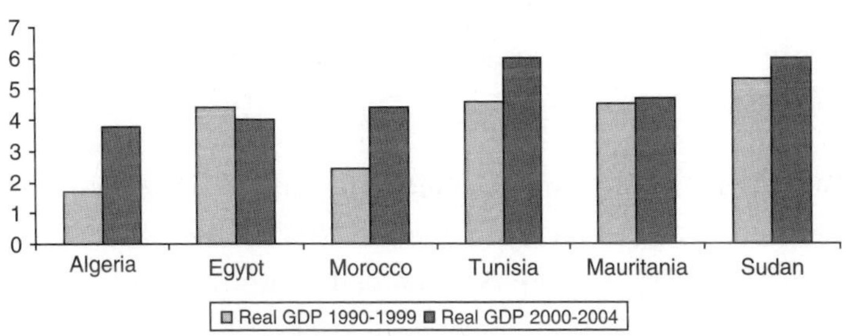

Figure 4.3: Average Annual Real GDP Growth in North Africa

Source: IMF (2006); and AfDB Statistics Department Database (2007).

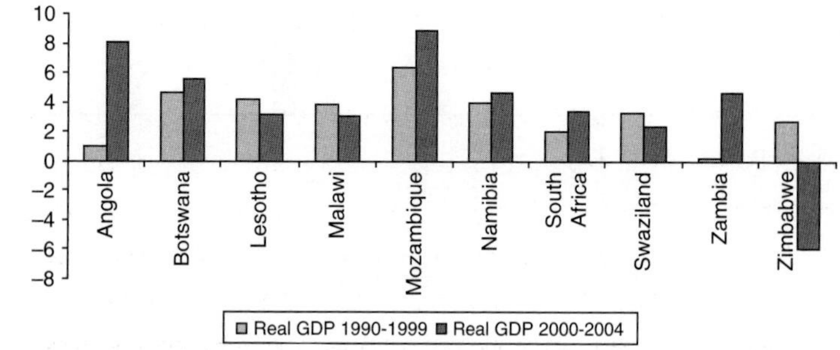

Figure 4.4: Average Annual Real GDP Growth in Southern Africa

Source: IMF (2006); and AfDB Statistics Department Database (2007).

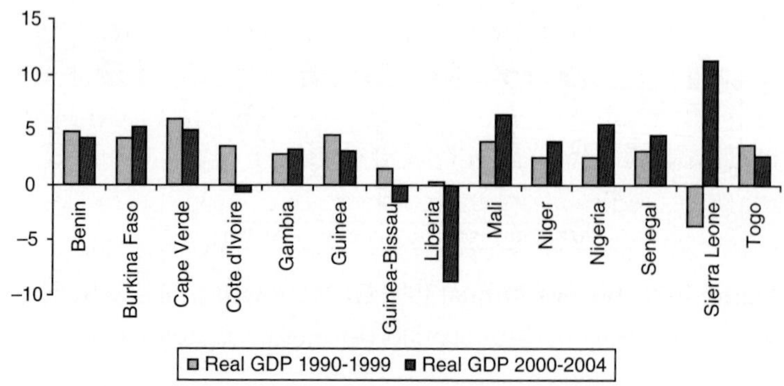

Figure 4.5: Average Annual Real GDP Growth in West Africa

Source: IMF (2006); and AfDB Statistics Department Database (2007).

the war in Iraq, policy developments in Venezuela, and conflicts in Nigeria contributed to the increase in oil prices in 2004. Suppliers have been attempting to comply with established quotas by increasing production. This was the case in 2004 when Angola, Chad, and Equatorial Guinea had their quotas raised four times. The main result of high oil prices in oil-exporting countries is increase in export revenues and hence higher economic growth. For example, in 2004 countries like Angola (Figure 4.4), Chad, Congo, Gabon, and Nigeria (Figure 4.5) recorded gains of more than 3 percent.

The impact of high oil trade can also be discussed from the perspective of terms of trade in African subregions, as shown in Figures 4.6 to 4.9. The figures suggest terms of trade gain during the 2000–2004 period. Net oil-exporting countries (such as Algeria, Egypt, and Tunisia) have shown steady positive terms of trade index.

GDP data actually supports the view that the aggregate output performance of most African countries has not been as seriously affected as expected by the impact of higher oil prices in the post-1999 period. This is in contrast with the expected downturn in aggregate output predicted by macroeconomic models. In fact, economic growth in the first half of this decade was far higher than the record in the 1990s (AfDB, 2005;

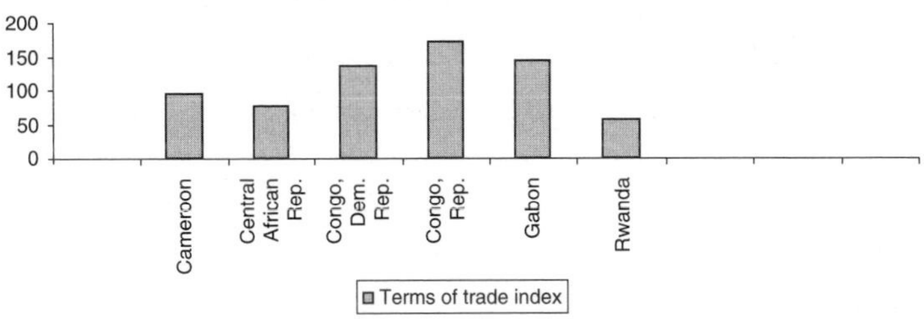

Figure 4.6: Terms of Trade in Central African Countries, 2000–2004

Source: IMF (2006); and AfDB Statistics Department Database (2007).

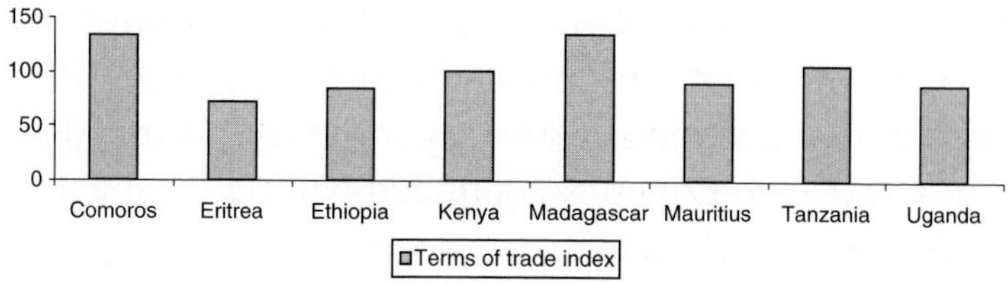

Figure 4.7: Terms of Trade in East African Countries, 2000–2004

Source: IMF (2006); and AfDB Statistics Department Database (2007).

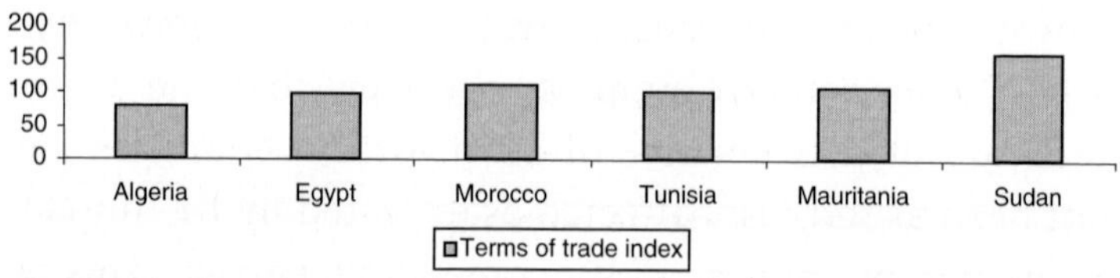

Figure 4.8: Terms of Trade in North African Countries, 2000–2004

Source: IMF (2006); and AfDB Statistics Department Database (2007).

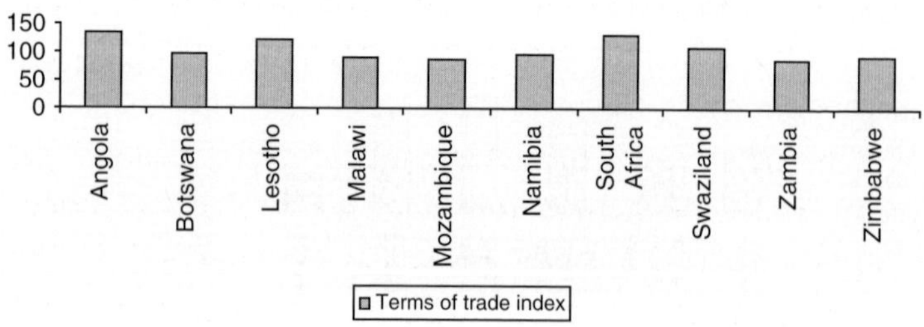

Figure 4.9: Terms of Trade in Southern African Countries, 2000–2004

Source: IMF (2006); and AfDB Statistics Department Database (2007).

World Bank, 2006*e*; IMF, 2006). Other macroeconomic variables such as investment, consumption, inflation, and fiscal balance also showed more robust performance. The available evidence thus suggests that the 2003–2005 oil price increase was not associated with any perceptible change in economic performance in net oil-importing countries unlike in the 1970s and the 1980s (Dudine et al., 2006). A similar conclusion was derived from a number of recent studies (World Bank, 2006*a*; 2006*c*; and 2006*d*), and earlier studies (ESMAP, 2005*a*; and 2005*b*) that quantified the magnitude of the oil price shocks relative to the size of the economy.

The impact studies tend to confirm that, based on data for 2004 and 2005, the slow down in GDP for non-oil exporters has been modest, and inflation has increased only slightly (Figure 4.10). These corroborative conclusions may just mean that the GDP and the terms of trade are too coarse as indicators of the real impact of high oil prices in African economies. African economies are very specific in the sense that they are dominated by informal economies, which are not adequately captured in macroeconomic indicators.

In practice, the situation is more complex because oil prices and GDP growth run both ways. High oil prices could dampen economic growth as experienced during earlier oil price shocks. A decline in economic growth would subsequently reduce demand, potentially leading to a large fall in world oil prices. During this latest relatively minor financial crisis, Brent oil prices fell below US$10 a barrel (December 1998).

4.1.3 *Impact of High Oil Prices on Domestic Prices*

In order to assess the real impact of high oil prices in African economies, it is important to understand that African economies are changing and significant improvements in terms of macroeconomic management are being reported, especially the very significant reduction in inflation (Figure 4.10).

The key policy response to the impact of high oil prices is the extent to which governments have passed on the price increases to consumers, or have moderated them with subsidies, tax reductions, or pressure on oil companies to hold down prices. In the countries that

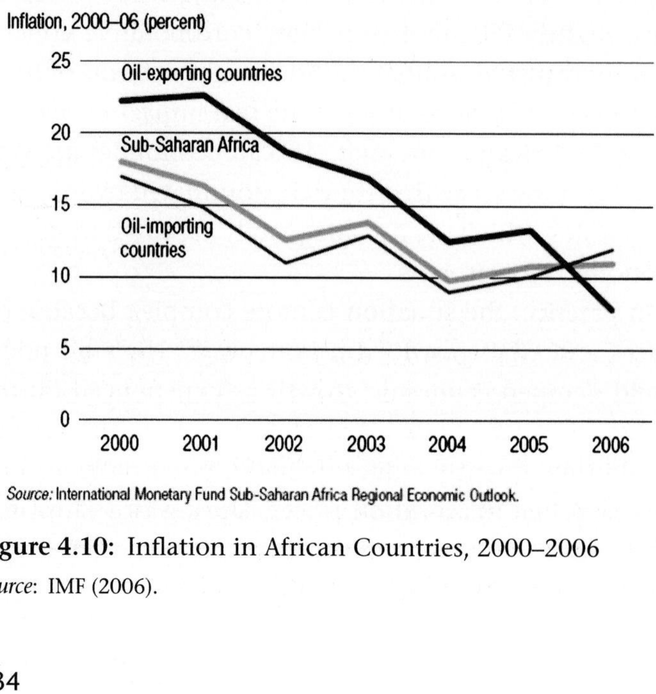

Figure 4.10: Inflation in African Countries, 2000–2006
Source: IMF (2006).

have been raising prices, and especially in those that have been passing on the full price increases to end-users, it is expected that some reduction in the demand for oil will occur. Where domestic prices have not fully reflected international price movements, the demand for oil products tends to stay more buoyant, and the potential negative effects on the balance of payments and GDP are larger. A central issue in the adjustment of the economy to any price rise that has occurred is thus the magnitude of the price elasticity of demand for oil products (ESMAP, 2006).

The extent of pass-through can be observed in Figure 4.11. It has been high in most countries; partial in many countries; and complete in others. A special case is Eritrea, where the increase in domestic prices was eight times the change in the international price of oil.

Substantially higher prices in recent years are illustrated in Figures 4.12 and 4.13. In fact, domestic

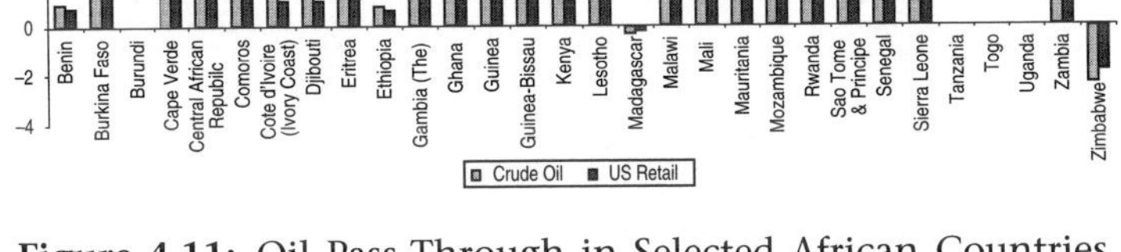

Figure 4.11: Oil Pass-Through in Selected African Countries, 2002–2005

Source: IMF (2006); and AfDB Statistics Department Database (2007).

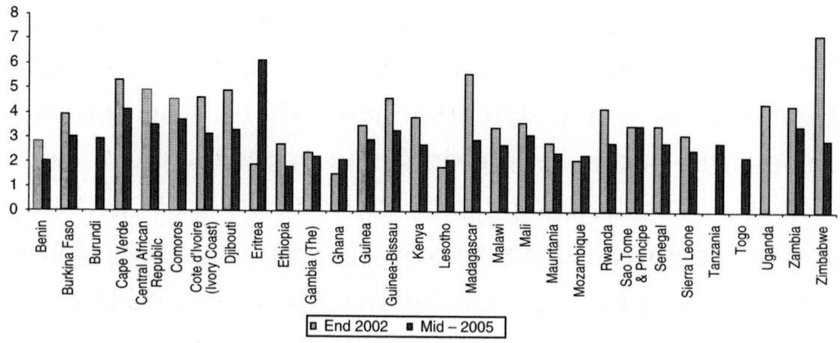

Figure 4.12: African Oil Product Prices Relative to Crude Oil Prices (Price in US Cents per Liter)

Source: IMF (2006); and AfDB Statistics Department Database (2007).

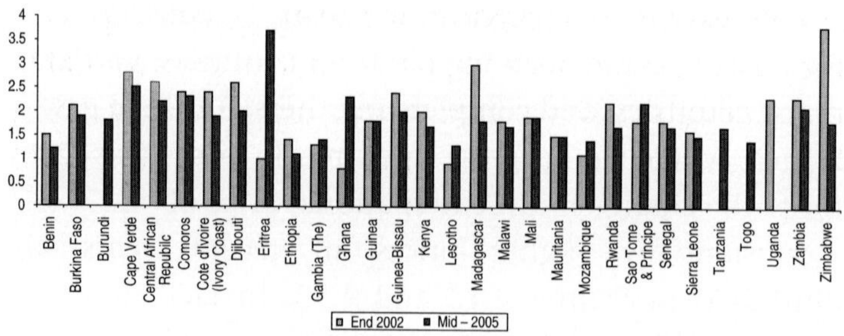

Figure 4.13: African Oil Product Prices Relative to US Retail Prices (Price in US Cents per Liter)

Source: IMF (2006); and AfDB Statistics Department Database (2007).

prices in most African countries, which are among the poorest countries in the world, are higher than in the United States.

4.1.4 *Social Impacts of High Oil Prices in Africa*

In African countries, the social impacts of high oil prices are direct and most outrageous in the poorest communities, which do not have electricity and use kerosene for illumination. It is also well known that increases in the prices of commodities that feature prominently in the consumption basket of poor households have severe consequences on their living standards and well-being. In other words, higher oil prices exacerbate the incidence and depth of poverty and highly distort income distribution structures.

For net oil exporters, higher oil prices are expected to be a blessing. However, key economic and social indicators for oil-exporting countries suggest that oil wealth has not been able to support sustained economic growth and development (see Chapter 3). Moreover, inequitable distribution of oil revenue among the population can fuel social tensions as has been witnessed in the case of the Niger Delta region.

4.2 AfDB Model of the Impact of High Oil Prices

4.2.1 *Overview of the Model*

The Research Department at the African Development Bank (AfDB) has developed a model to quantify the impact of high oil prices on oil-importing and oil-exporting African economies.[5] The analysis is based on a dynamic stochastic general-equilibrium model of a small open economy. The model is rigorously

[5] "The Impact of High Oil Prices on African Economies," Working Paper Series (WPS) No. 93, December 2007, African Development Bank Group, Tunis.

micro-founded in the sense that agents are fully optimizing and form their expectations in a rational manner. This modeling approach has several advantages. First, the inter-temporal nature of the model permits a study of the effects of transitory as well as persistent oil price shocks. Second, because the time path of aggregate variables is determined by the optimizing behavior of economic agents, the model is robust to the *Lucas critique*, and, therefore, appropriate for policy analysis. Third, this approach allows proper welfare analysis as it provides an explicit account of households' preferences. Fourth, the general-equilibrium perspective ensures internal consistency and, more importantly, allows the study of the effect of oil prices without making arbitrary assumptions about what is exogenous and what is not.

Owing to these advantages, the proposed model supersedes existing naive data-based models, reduced-form models, and static computable general-equilibrium (CGE) models. Data-based models are useful to characterize the statistical relationships between economic variables and to establish relevant stylized facts; however, these models lack economic content and do not reveal mechanisms through which shocks propagate. Reduced-form models are often concise and easy to solve, but the aggregate relationships are usually not derived from first principles and the model parameters are not always invariant to shifts in policy regimes. The same criticism applies to CGE models, which, in addition, often ignore inter-temporal considerations.

The proposed model belongs to the class of new open-economy macroeconomic models, which have become the main tool used in modern international macroeconomics and are increasingly sought by international organizations and Central Banks around the world. It is a state-of-the-art model that uses cutting-edge techniques to address positive and normative issues related to the effects of oil price shocks, but which can easily be extended to examine broader questions of relevance to the AfDB.

The model is one of a small open economy and is adapted specifically to the context of oil and African economies. It is sufficiently flexible to represent virtually any African country. In particular, it is configured to describe oil importers and oil exporters, credit-constrained economies and those that have access to international financial markets, and countries with flexible, managed, and fixed exchange rate regimes. The economy consists of households, firms, a government, and a monetary authority. There are four types of goods: final goods, composite non-oil goods, oil, and intermediate goods.

4.2.2 *Results*

This section discusses the impact of a doubling in world oil prices on main macroeconomic variables both in the case of a median oil-importing economy and a median oil-exporting economy. The variables of interest are output, consumption, inflation, real exchange rate, government budget deficit, and foreign debt. The simulations are performed under both a fixed exchange rate regime and a managed float. For each case, two different scenarios are considered: complete and zero pass-through. In all simulations, the oil price shock is assumed to be persistent, with a first-order autocorrelation coefficient of 0.85, as estimated from the data. This assumption is consistent with the view that the expected durability of high oil demand from East Asia (especially China) is sustaining the market expectations that oil prices will remain high.

4.2.2.1 MEDIAN OIL-IMPORTING ECONOMY

This economy is calibrated such that oil imports represent roughly 13 percent of total imports and 5 percent of total GDP in the steady state. Simulation results for this case are shown in Tables 4.2 and 4.3. The main conclusions are as follows:

- Under fixed exchange rates and complete pass-through, a doubling in the world price of oil leads to a decline in output and consumption, a slight increase in inflation, a small appreciation of the real exchange rate, and moderate changes in public and foreign borrowing. The output loss is about 6 percent during the first year, while the cumulative loss is about 24 percent during the five years following the shock. For consumption, the corresponding numbers are 5 and 19 percent, approximately.

Table 4.2: Effects of a 100% Increase in the Price of Oil (Net Oil-Importing Country, Fixed Exchange Rate Regime)

	Impact effect (1 year) %	Cumulative effect (5 years) %
Output		
Complete pass-through	−6	−24
Zero pass-through	−1	−5
Consumption		
Complete pass-through	−5	−19
Zero pass-through	−6	−25
Investment		
Complete pass-through	−11	−39
Zero pass-through	−7	−25
Inflation		
Complete pass-through	2	1
Zero pass-through	−4	−4
Real exchange rate		
Complete pass-through	−2	−7
Zero pass-through	4	22
Budget deficit		
Complete pass-through	4	7
Zero pass-through	31	45
Foreign debt		
Complete pass-through	−1	2
Zero pass-through	9	11

Note: Budget deficit in percentage of steady-state output.

- The drop in output and consumption is attributed to a combination of two effects of high oil prices: a direct income effect, through the resource constraint, and a direct effect on production, through higher costs of inputs. The former decreases consumption and increases labor supply. The latter decreases demand for non-oil inputs and, by extension, demand for labor and capital. The net effect on

Table 4.3: Effects of a 100% Increase in the Price of Oil (Net Oil-Importing Country, Managed Floating)

	Impact effect (1 year) %	**Cumulative effect (5 years) %**
Output		
Complete pass-through	−6	−23
Zero pass-through	2	−1
Consumption		
Complete pass-through	−4	−18
Zero pass-through	−5	−25
Investment		
Complete pass-through	−10	−38
Zero pass-through	−1	−21
Inflation		
Complete pass-through	5	4
Zero pass-through	4	5
Real exchange rate		
Complete pass-through	−1	−5
Zero pass-through	9	30
Budget deficit		
Complete pass-through	0	−1
Zero pass-through	6	20
Foreign debt		
Complete pass-through	1	2
Zero pass-through	16	12

Note: Budget deficit in percentage of steady-state output.

hours worked is ambiguous, but labor income and investment unambiguously fall (due to lower marginal productivity of labor and capital). The resulting reduction in households' disposable income further decreases consumption and output.

- The increase in inflation is due to the fact that the domestic price of oil enters the aggregate price index, and since there is complete pass-through, oil-price inflation contributes to core inflation. The higher inflation explains the appreciation of the real exchange rate (since the nominal exchange rate is fixed).

- Under zero pass-through, the increase in the price of oil still leads to a decline in output and consumption, but the magnitude of the effects differs significantly compared with the complete pass-through case. The decline in output during the first year is less than 1 percent and the cumulative loss during the five years following the shock is roughly 5 percent. Hence, by choosing a zero pass-through, the government shields the production sector of the economy, thus minimizing the output loss. However, the cost of this intervention is a dramatic deterioration in the budget deficit (31 percent during the first year and 45 percent after five years), and, most importantly, a large decline in consumption, which drops by more than 6 percent during the first year and 25 percent after five years.
- Under zero pass-through, there is a decrease in inflation, which translates into a real exchange rate depreciation of roughly 4 percent in the first year and 22 percent after five years.
- Under a managed floating exchange rate regime, the nominal exchange rate is, to a certain extent, free to adjust, thereby acting as a shock absorber. In principle, therefore, the adverse effects of high oil prices should be less severe compared to the case

with fixed exchange rates. A comparison of Tables 4.3 and 4.4 confirms this intuition. Under complete pass-through, however, there are only minor differences in the response of output, consumption, inflation, and, to a lesser extent, foreign debt across the two regimes.[6] The gain from letting the nominal exchange rate float is much more apparent under zero pass-through. For example, output initially increases by almost 2 percent (as opposed to a decline of 1 percent) following the rise in the price of oil, and the cumulative loss after five years is barely over 1 percent (as opposed to a loss of 5 percent). This smaller output loss is due to the larger depreciation of the real exchange rate relative to the case with pegged nominal exchange rates.

4.2.2.2 MEDIAN OIL-EXPORTING ECONOMY

This economy is calibrated such that oil exports represent roughly 88 percent of total exports and 35 percent of total GDP in the steady state. Simulation results for this case are shown in Tables 4.4 and 4.5. The main conclusions are the following:

- Under fixed exchange rates and complete pass-through, a doubling in the world price of oil leads to a 9 percent increase in output, a 42 percent increase in consumption, a 9 percent increase in inflation, a

[6] The only notable difference across the two regimes is the response of the budget deficit, which deteriorates under the peg one, but slightly improves under managed floating.

Table 4.4: Effects of a 100% Increase in the Price of Oil (Net Oil-Exporting Country, Fixed Exchange Rate Regime)

	Impact effect (1 year) %	Cumulative effect (5 years) %
Output		
Complete pass-through	9	53
Zero pass-through	10	56
Consumption		
Complete pass-through	42	152
Zero pass-through	41	149
Investment		
Complete pass-through	16	62
Zero pass-through	16	62
Inflation		
Complete pass-through	9	15
Zero pass-through	6	14
Real exchange rate		
Complete pass-through	−9	−71
Zero pass-through	−7	−63
Budget deficit		
Complete pass-through	−114	−147
Zero pass-through	−108	−139
Foreign debt		
Complete pass-through	−33	−47
Zero pass-through	−30	−45

Note: Budget deficit in percentage of steady-state output.

9 percent real appreciation, a 114 percent reduction in the budget deficit, and a 33 percent reduction in foreign debt during the first year. The magnitudes of the cumulative effects after five years indicate that the adjustment of output, the real exchange rate, and foreign debt is non-monotonic.[7]

[7] For example, the model predicts that the response of output to the 100 percent increase in the price of oil is hump-shaped, attaining its peak of 16 percent during the third year after the shock.

Table 4.5: Effects of a 100% Increase in the Price of Oil (Net Oil-Exporting Country, Managed Floating)

	Impact effect (1 year) %	Cumulative effect (5 years) %
Output		
Complete pass-through	4	25
Zero pass-through	4	27
Consumption		
Complete pass-through	16	75
Zero pass-through	16	76
Investment		
Complete pass-through	3	22
Zero pass-through	4	23
Inflation		
Complete pass-through	−13	−12
Zero pass-through	−14	−13
Real exchange rate		
Complete pass-through	−38	−136
Zero pass-through	−36	−130
Budget deficit		
Complete pass-through	−7	−24
Zero pass-through	−6	−23
Foreign debt		
Complete pass-through	−55	−39
Zero pass-through	−53	−38

Note: Budget deficit in percentage of steady-state output.

- The increase in the price of oil generates a positive income effect, through the resource constraint, which increases consumption. This rise in consumption translates into higher demand for the final good, which more than offsets the negative effect of the higher price of oil. As a result, the demand for oil and non-oil inputs increases (due to their complementarity), thereby raising the demand for labor and capital. The resulting increase in labor demand and investment further boosts the demand for the final good, and, therefore, output.

- Under zero pass-through, there is a slightly larger increase in output, a lower inflation, and a smaller appreciation of the real exchange rate compared to the case with complete pass-through. However, this "gain" comes at the expense of a (marginally) smaller increase in consumption and a smaller improvement in the budget deficit.
- Under managed floating, the output and consumption gains induced by the increase in the price of oil are smaller than under fixed exchange rates. This result is mainly due to the larger appreciation of the real exchange rate under the former regime. The smaller increase in consumption implies that the budget deficit narrows less than under the fixed exchange rate regime.
- Under a managed float, the effects of an increase in the price of oil under complete and zero pass-through are strikingly similar.

For the sake of comparison, Table 4.5 is generated with the same parameters as Table 4.4, except for the exchange rate regime. The deflation reported in Table 4.5 arises because the appreciation of the local currency makes imported goods cheaper. Reducing the share of imported goods in the economy and running the simulation over would give positive inflation.

4.2.3 *Policy Implications*

4.2.3.1 GOVERNMENT INTERVENTION

The above analysis suggests that through local currency pricing (buying at world market prices, but selling at local prices), the government can cushion the economy from the adverse effects of oil price shocks in oil-importing countries. However, this policy amplifies a consumption loss and aggravates the government's budget deficit. Hence, the answer to the question of whether a government should intervene or not depends on its implicit objective function. To the extent that the government is concerned with stabilizing output, local currency pricing proves to be the optimal policy. Alternatively, if the government is a benevolent social planner, then laissez-faire and pass-through is likely to be the welfare-maximizing policy. For oil-exporting countries, government intervention does not seem to affect in a substantive way the outcome of the economy, especially in the case of a managed floating. This observation implies that both intervention and laissez-faire could be acceptable policy choices in these countries.

4.2.3.2 FOREIGN AID

Can foreign aid help African oil-importing countries cope with high oil prices? Are the required amounts prohibitive? Table 4.6 shows the permanent level of overseas development assistance (in percentage of steady-state output) that is required to completely offset the initial output loss associated with a persistent 100 percent

Table 4.6: ODA Needed to Offset Output Loss in the First Year (% of Steady-State Output)

	Fixed exchange rate regime	Managed floating
Complete pass-through	1.60%	1.98%
Zero pass-through	0.23%	—

Note: ODA: Official Development Assistance.

increase in the price of oil. The table shows that the largest amount of foreign aid needed is less than 2 percent of steady-state output. This amount is clearly non-prohibitive (foreign aid in a number of African countries represents more than 5 percent of GDP), implying that there is scope for international-community actions to help debt-burdened African economies mitigate the adverse effects of high oil prices.

4.2.4 *Summary of Results*

High oil prices can have very harmful effects on the economy of African oil-importing countries, especially those that are heavily debt-burdened. They lead to a decrease in output and consumption, and to a worsening of the net foreign asset position. For the median oil-importing country, the five-year cumulative output loss resulting from a doubling in the price of oil can be as large as 23 percent under a fixed exchange rate regime, as per the model applied. However, this recessionary effect can be cushioned through government intervention in the form of limited pass-through of the oil price increase or through foreign aid. In this regard, the model can be used to determine the optimal degree of intervention by the government and donors given their objective functions.

For the median oil-exporting country, the five-year cumulative increase in output associated with a doubling in the price of oil exceeds 60 percent, regardless of the exchange rate regime under which the country operates. However, this scenario is accompanied by a sharp appreciation in the real exchange rate, which may hinder the competitiveness of the country. It is therefore important that accompanying measures be taken to offset the exchange rate appreciation effect on non-oil sectors.

It should be emphasized, however, that while the analysis above focuses on "median" countries, there is, in fact, a great deal of heterogeneity within the groups of oil-importing countries and oil-exporting countries. This means that the effects of oil price shocks can differ dramatically from one country to another. As stated above, however, the proposed model can be configured to represent any of these countries.

An important question that the model does not address is the effect of high oil prices on poverty, which is a crucial dimension in the African context. The model can, in principle, be extended to capture this feature by allowing for heterogeneity across households and by assuming that some of them have liquidity constraints. The model can also be extended to include other types of shocks, such as productivity shocks, monetary-policy shocks, and world-interest-rate shocks.

4.3 Coping with High Oil Prices

Africa as a whole has in recent years been enjoying gains from macroeconomic management, improved political stability and governance, easing of regional conflicts, strengthening of regional economic blocks, and increased agricultural, mineral, and oil production. Before the financial crisis, many African economies were moving towards fast and steady economic growth. Their performance over 1995–2007 reversed the collapses over 1975–85 and the stagnations over 1985–95. And, for the first time in three decades, African economies were growing at the same rate as the rest of the world economies. Average growth in the Sub-Saharan economies was 5.4 percent in 2005 and 2006. However, sustaining high growth rates has been a chronic challenge for African countries as they confront shocks including high oil prices.

Coping with high oil prices requires a set of measures to maximize the positive impacts and mitigate the negative impacts. The negative impacts are more intense for net oil importers, but can (as described in earlier sections) be mitigated through government initiatives or through foreign aid. Specifically, some of the measures that governments can implement to cope with high oil prices include the following (ESMAP, 2006):

- Price-based policies (comprising full passing on of price increases and subsidies of end-user prices;
- Policies to reduce the cost of supply (hedging, economies of scale, security of stocks);
- Quantity-based policies;
- Diversification into non-petroleum sources of energy (natural gas, renewable power sources, biofuels);
- Increasing domestic supply; and
- Domestic policies (for example, winning public buy-in of government actions).

4.3.1 *Price-Based Policies*

4.3.1.1 FULL PASSING ON OF PRICE INCREASES

In a completely deregulated market, price increases are passed on fully to consumers. Linking domestic prices to international prices in a pricing formula attempts to mimic a deregulated market (Research Division, 2007). A formula-based pricing mechanism raises the question of the frequency of adjustment and the time period over which the reference price is averaged. Given the volatility of world product prices, in countries where governments wish to exercise some measure of control over pricing, it may be reasonable to take a moving average of actual prices spread over a period of more than one month. However, less frequent price adjustments based on a moving average of several months will lead to a temporary need to finance the difference between actual prices and smoothed recommended prices and possibly large adjustments when price adjustments are finally made. In these cases, there is the risk of allowing the formula price to diverge more sharply from the actual international price, with the consequent burden on financing. Less frequent price changes run the risk of needing to increase the formula-based price at a time when current market prices are decreasing, thus adding to consumer confusion and antagonism.

Adjusting formula prices very frequently, whatever the length of moving average chosen, is likely to track prices more closely, but it does impose increased administration costs for the process.

Another option to solve the problem would be to change prices when international (spot) prices change by more than a certain percentage in local currency. Retail prices in this case are adjusted as a result of foreign exchange fluctuations, international oil price movement, or both. The larger the minimum percentage change set to trigger an adjustment, the less frequently prices will be changed, but the larger the change. This mechanism shields consumers from small, frequent price changes, but does not achieve price smoothing in the face of large fluctuations (large price shocks outside of the price band are instantly passed on to consumers).

Petroleum product price stabilization funds have been used in some countries to manage revenue shortfalls in times of rising oil prices. However, a fund that has not accumulated large funds before the recent oil price increase would be unable to smooth prices without an initial transfer from the government.

A related policy is the establishment of a strategic stock, where the government acquires and stores crude oil or products at a time of lower prices, to be released when there is an actual supply shortage or a spike in world market prices. The difference in the prices experienced may be sufficiently large to cover the costs of financing the purchase and storing the oil. A further development for smoothing prices to final users would be for the appropriate government agency to hedge future product prices.

Governments that consider a full pass-through of prices to consumers are more concerned with the effects on the rate of inflation as well as on fuel purchasers. If the price rise were to become embedded into the core rate of inflation, macroeconomic policies would be needed to control the inflation created. This usually calls for tighter control over monetary policy and some form of wage restraint. These issues will be even more severe for a government that subsidizes product prices and is considering removing the subsidies and setting prices at the international market level.

4.3.1.2 SUBSIDIES OF END-USER PRICES

Subsidies can be regarded as any government intervention that lowers the price of a fuel below its economic opportunity cost. These relate to the structure of costs, and to whether or not the country is an oil producer and refiner (Research Division, 2007). There are, indeed, various points along the supply chain where subsidies can be introduced.[8]

Even before the recent acceleration in oil price increases, a number of countries were subsidizing some or all petroleum product prices for final users. Because the burden of a complete subsidy to cover all oil price rises since 2004 has been higher and higher, governments have been anxious to find schemes that subsidize only part of the price increase, largely by providing differential subsidies on the various products and to different consumer groups.

The amount of the subsidy and its magnitude for different products has to be decided by individual governments on the strength of country-based analysis. Government decisions on subsidies thus depend on various factors that include share distribution of the different petroleum products in their total demand, specific use, economic groups within society, but, first and foremost, the finance ability (funds) available to finance on a sustained basis. For example, in a country where gasoline imports are significant and gasoline is known to be consumed mainly by private vehicles owned by better-off households, the government may decide not to subsidize gasoline imports. In contrast, if diesel is used for agriculture and long distance transport of goods and people, usually benefiting a wider population and fueling the domestic economy, the government may

[8] ESMAP (2006) provides a framework and several relevant examples.

decide to subsidize—totally or partially—the impacts of high prices on diesel imports. The government may also decide to subsidize the import of kerosene, known to be largely used by low-income households, in a desire to protect this segment of the society.

Subsidies are usually enacted through controlled final prices, possibly fixed by a formula that passes on only a fraction of international price increases. Where petroleum product sales are in the hands of a state body, it may attempt to finance its shortfall through borrowing from the domestic banking system, but such a policy may not be sustainable.

4.3.2 *Policies to Reduce the Cost of Supply*

If the cost of supply can be lowered in a market-based way and the cost reductions passed on to consumers, end-user prices would not rise as much. Governments have sought to reduce costs or margins in the downstream petroleum product sector in a number of ways, depending on the government's ability to finance (Research Division, 2007).

4.3.2.1 CREATING ECONOMIES OF SCALE

In smaller economies, each supplier may be too small to obtain the lowest possible price for imported products. Accordingly, some countries have established bulk purchasing agencies (for example, in Mozambique all companies request the required quantity in a specific period; the government body procures the products and after sealing the deal collects the equivalent value from each company) or agreements between companies, so that economies of scale can be obtained through single purchasing arrangements. The centralized procurement system allows governments to control the minimum and maximum prices according to an added transport differential and allowed maximum mark-up.

4.3.2.2 HEDGING PRODUCT PURCHASES

Commercial companies, particularly those that are subsidiaries of large international companies, may well have considerable experience to draw upon in hedging product purchases. This can help smooth out variations in the costs of purchasing products, but cannot be expected to lower the average price paid over a lengthy period. Governments may wish to be informed of purchasing costs, especially in a regulated system, but are unlikely to wish to mandate private sector hedging.

4.3.2.3 ENSURING SECURITY OF STOCKS

Many governments require product marketers to carry a certain amount of security stocks (measured in terms of a given number of days of normal consumption). This provides a buffer against disruptions in supply caused by possible delays in delivery. As a precaution against more erratic supply conditions that appear to be associated with price spikes in the market, governments can increase this requirement, or establish their own security stocks. A mechanism for bearing the cost of the stocks has to be agreed. These costs include both the interest on working capital tied up, as well as the costs of the storage facilities themselves.

It is possible to use these stocks to smooth out large swings in prices, but this strategy is likely to be effective only in short-lasting price hikes; in this case, purchasing of supplies can be delayed until the price returns to normal. Security stocks are more likely to be used when there is physical disruption, and supplies are actually disrupted. Landlocked countries, which have very limited sources of supply, are usually the most vulnerable to such disruptions and have the greatest need of some security cover from extra stocks.

4.3.2.4 INTRODUCING COMPETITION FOR THE MARKET

If the government wishes to maintain ownership of the downstream sector but is prepared to entrust its management to a private sector company, it can use a "management contract" approach in which private sector operators bid for the right to operate the market. The bids can be tied to the price formula whereby the best bid is the one that will deliver the lowest price. These schemes usually limit the duration of the contract and give the government the right to appoint another operator after a period of time. This approach may be especially relevant for supplying remote areas. Some governments impose pan-territorial pricing to protect consumers in remote areas that are expensive to supply. Instead of imposing pan-territorial pricing, it may be more cost-effective to liberalize pricing and offer protection to consumers in remote areas by means of, for example, competition for an exclusive right to supply certain regions at the lowest subsidy. The supervision and control of compliance with such measures will impose a big challenge to governments.

4.3.2.5 INCREASING THE NUMBER OF SUPPLIERS IN THE MARKET

Where the downstream market is already privatized, governments can still increase the number of suppliers by placing legal limits on market shares at various stages of the downstream. This policy requires existing firms to divest some of their assets. Policies toward the number of competing suppliers in a market depend on the size of that market. For a large market, each of a number of suppliers can achieve a large enough turnover to enjoy the available economies of scale. For a small market, however, dilution of market share can result in a loss of economies of scale that would negate any cost reductions from a more competitive environment. Countries that have permitted a very large number of firms at retail tend to find that these are not necessarily efficient and that they are difficult to monitor for pricing, quantity, and quality purposes.

4.3.3 *Quantity-Based Policies*

Many countries have considered quantity-based policies with respect to a wider issue of security of supply for oil and petroleum products, of which higher prices are a subset. These policies come in two principal forms:

- Policies that effectively ration the purchase of oil products; and
- Policies that effectively ration activities that intensively use oil products.

Schemes for rationing the purchase of petroleum products that are sold at a large number of outlets, as is typical for transportation fuels and for kerosene, cannot rely on the seller to impose a ration based on previous purchases by an individual consumer. These can be achieved by setting differentiated maximum limits per week for private and commercial vehicles. An effective ration system serves to cut demand, but rationing a commodity almost inevitably leads to a black market.

4.3.4 *Increasing Domestic Supply*

An increasing number of countries are promoting oil and gas development. With rising oil prices, the economics of exploration, development, and production have changed. Even in areas where past exploration did not yield promising results, there are hopes that newer exploration technologies might lead to commercial discoveries. This is the case of Africa, particularly in the

Indian Ocean shoreline, where exploration efforts have identified potential reserves in Madagascar, South Africa, Mozambique, Tanzania, Kenya, and Uganda as new ventures for oil companies.

A variation on increasing the domestic supply of oil and gas is increasing domestic refining capacity. Several arguments have been put forward in favor of enhancing refining capacity (ESMAP, 2006):

- Although a fundamental disagreement persists over the causes of higher oil prices, some industry analysts have argued that one reason is tight refining capacity. Increasing refining capacity would address this bottleneck directly and would, in time, help lower world oil prices.
- Sufficient domestic refining capacity could enhance security of supply of refined products.
- Refining margins have widened in recent years, especially for complex refineries with conversion capacity that can convert lower-priced heavy crude into light products (LPG, gasoline, kerosene, and diesel). Domestic refineries can capture higher margins and pass benefits on to the economy through taxes paid to the government. It is important to note, however, that refining margins fluctuate widely.

Significant economies of scale are an important asset in refining. Many refineries in developing countries are doing poorly even in the current climate of high refining margins because they are too small and lack sufficient conversion capacity. Small-scale refineries are unable to capture the economies of scale that are necessary to bring costs to a level that can compete with international prices.

The other important factor in the refining industry is the capital costs of constructing competitive refineries, which are very large and would constitute a substantial fiscal burden for any government that chooses to support a state-owned oil company that wishes to enter such a market.

The refinery industry requires large quantities of crude oil, which means that the best location of a refinery is close to a deep sea port. African countries are already thinking about increasing the refinery capacity in the region. Mozambique, for example, is setting up a US$ 7 billion investment refinery in the Nacala deep sea port.

4.3.5 *Diversification into Non-Petroleum Sources of Energy*

Reliance on oil can be reduced by diversifying into non-petroleum sources of energy. The most common alternatives are natural gas and renewable sources of electricity such as hydro, geothermal, solar, and wind. Biofuels are attracting growing attention as a substitute for liquid transportation fuels. Some consumers are forced back to use more solid biomass, with damaging public health and environmental effects.

There are economies of scale in laying down pipelines. A pipeline network is economically viable if there are large consumers (such as power plants and large industrial operations). Natural gas prices have risen markedly in recent years, broadly tracking world oil price increases.

A policy of promoting the blending of a certain proportion of ethanol into gasoline or biodiesel into diesel will reduce the total use of petroleum transportation fuels. This policy may help reduce the fuel import bill in the short run if a country can import at prices lower than those of petroleum fuels, or if there is surplus feedstock and production capacity.

The use of biofuels supports several major energy policy objectives:

- Energy security: Biofuels can readily displace petroleum fuels, and, in many countries, can provide a domestic rather than imported source of transport fuel. Even if imported, ethanol and biodiesel will likely come from regions other than those producing petroleum (for example, Latin America rather than the Middle East), creating a much broader global diversification of supply sources of energy for transport;
- Environment: Biofuels are generally more climate-friendly than petroleum fuels, with lower emissions of CO_2 and other greenhouse gases;
- Fuel quality: Refiners and car manufacturers have become very interested in the benefits of ethanol in order to boost fuel octane, especially where other

potential octane enhancers, such as MTBE (methyl-tertio-butyl-ether), are discouraged or prohibited;

- Sustainability in transportation is derived from renewable energy.

4.3.6 *Other Domestic Policies*

A number of governments that have attempted to eliminate subsidies on petroleum products by imposing large price increases have faced strong reaction from civil society—trade unions and non-affiliated groups (the riots in Mozambique in February 2008, for example, were due to a more than 50 percent rise in public transport costs as a consequence of high oil prices). Political opposition parties have also built on hostility to these price increases to encourage demonstrations and protests, which may be opportunistic, but which may also be aimed at wider dissatisfaction with the government.

Many governments have deliberately adopted policies that reduce the burden of higher prices on certain groups within society. Policies that are transparent in formulation and implementation are more likely to be accepted than those that are not. Transparency entails both making information widely available to the public and selecting measures that are easily verifiable.

4.4 Summary and Conclusion

In summary, the role of economic policy—at both the micro and macro levels—in mitigating the potential adverse economic impact of higher oil prices cannot be overemphasized. The impact of changes in oil and gas prices will remain an important issue in the debate on Africa's social, economic, and political development. Sustained development and substantial improvement in the living standards of the people depend fundamentally on secure energy resources, and, in the short term, at least, adequate supplies of oil and gas at affordable prices.

A fundamental disagreement persists about the causes of higher oil prices, and some industry analysts have argued that one of the causes is tight refining capacity. Increasing refining capacity would address this bottleneck directly and would, in time, help lower world oil prices. Having sufficient domestic refining capacity could enhance security of supply of refined products.

Regardless of the reasons, the economic impact of higher oil and gas prices is devastating and includes reduction in economic growth and in household well-being as a result of real income decline, macroeconomic disequilibrium, and inflation. The adverse economic and social effects of higher oil prices are often greater in net oil-importing economies. The challenge is to design robust policies that will mitigate these economic effects. Notably, the emerging macroeconomic evidence suggests that net oil-importing and more oil shock-prone African countries have a greater capacity for coping with recent oil price shocks.

Higher crude oil prices are often associated with greater profits from upstream oil activities by international and national oil companies. Where these are vertically integrated into refining and selling petroleum products, there is a widespread demand from consumers that some of the profit be returned to them in the form of lower prices. For companies that are not involved in upstream oil production, a policy that limits price increases may not be sustainable, unless there are inefficiencies in supply that they can remove under the pressure of reduced profits.

In several countries, especially in net oil-exporting countries, many governments are not willing to fully accept the tradeoffs between the distributive effect and efficiency considerations associated with competitive oil markets because of popular opposition to substantial upward adjustment in the pump prices of oil products. This makes the product essentially a political good in a number of countries. The price differential in oil- and non-oil-exporting countries reflects the political content of oil product pricing in the region—hence the appropriate price of oil products remains a controversial issue in many countries in Africa. Administered pricing is common in the region. The size of the pass-through is also contentious. In some countries, the pass-through is full, while in others, it is partial. Obviously, in dealing with the demand and supply dimensions of expended energy access, oil and gas pricing, the size of the pass-through and oil and gas market liberalization are essential elements in the energy equation.

Coping mechanisms depend on country-specific circumstances, especially the country's ability to cushion the effects of high oil prices on vulnerable groups.

Coping policy mechanisms could include issues related to strategic reserves and inventories, managed as a defensive stockpile. Oil price shocks in many African countries can be partially mitigated given innovative oil inventory management strategies. Yet, most of the countries do not have, or cannot afford to have, any sizable oil inventory.

In some countries, governments have ensured that prices charged by companies do not rise as fast as the costs of products on the world market. The policies used to achieve this have been either mandatory, where price caps have been placed on petroleum products, or exhortatory, whereby governments have encouraged such action. In either case, if there is no offsetting compensation from the government, the company bears the burden of the price rise.

The resilience of economies in coping with the challenges of oil price shocks is a function of their conditions prior to and during the shocks. How to finance higher oil prices, especially in developing countries, remains a contentious issue. Increased transparency will better inform consumers and policy-makers in the executive and legislative branches about factors that affect the level and volatility of prices for refined petroleum products.

African oil exporters need to establish and enforce prudent fiscal rules to smooth surplus export receipts over time, invest them for future growth, and minimize wasteful spending.

A key challenge for policy-makers in Africa is designing the optimal policy mix that would help African countries cope efficiently with the economic and social costs of the external shocks accompanying higher oil and gas prices. In more recent times, macroeconomic policies have been more robust to the oil price dynamics. However, what has emerged as consensus among analysts is that micro- and macroeconomic policies can help mitigate the potential adverse economic impact of high oil prices.

5

Challenges and Opportunities

5.1 Making Oil and Gas Wealth Work for the Poor

This chapter examines the challenges and opportunities for turning oil and gas resources into economic wealth to alleviate poverty in Africa. The discussion focuses on providing a platform that will allow for pursuit of the *best opportunities in the future* to fully harness wealth from oil and gas resources.

At the regional level, the energy demand-supply balance situation in Africa is positive: Given the region's total energy resource endowment, substantially higher oil and gas consumption can adequately be met with regional supplies—with ample surplus still maintained for export. Despite this seemingly optimistic outlook, the country and sub-regional realities reveal multidimensional issues and challenges confronting expanded access to oil and gas in Africa (AU, 2007).

Addressing the issue of volatile and high oil prices calls for a holistic approach; both the short- and long-term dimensions of the likely economic and social impacts must be addressed. The short-term considerations include how to manage the immediate impact of higher oil prices. This has been the preoccupation of several net oil-importing countries in the region. However, exploring the short-term issues without adequate consideration of medium- and longer-term issues puts these countries at risk of being perpetually trapped in a low-level energy and low-growth equilibrium. The other considerations are medium- to long-term in nature and concern economic, institutional, and organizational considerations required to pursue the important goal of achieving expanded access to oil, gas, and other energy forms in African countries (AU, 2007). The following sections address these issues in greater detail.

5.2 Oil and Gas Wealth and the Enabling Environment

Making rational choices about oil and gas resources is central to making natural wealth work for the poor and to maximizing the contribution that the resource endowment can make to social and economic development in Africa. Three key elements—(1) environment and nature (oil and gas resources); (2) wealth; and (3) the enabling environment (governance, institutions, and power structure)—and their relations provide a useful

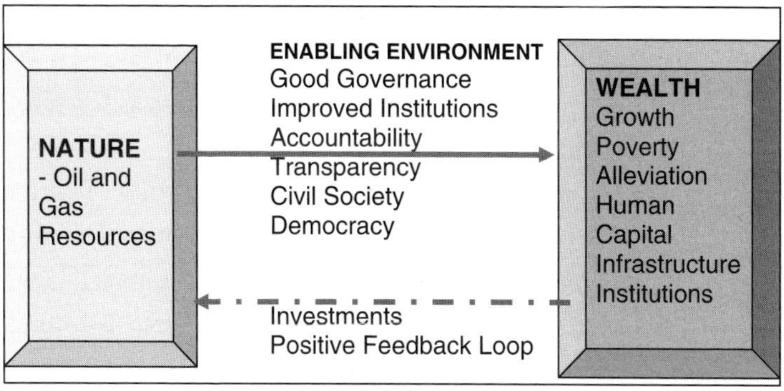

Figure 5.1: Process of Transforming Oil and Gas Resources into Wealth

Source: Authors.

framework for discussing how wealth from oil and gas resources can enhance development and chart a path forward (see Figure 5.1). Experience in Africa and elsewhere demonstrates that policies and investments that recognize and integrate these three elements yield positive development outcomes (see, for example, USAID et al., 2002; and AfDB, 2007*a*).

Figure 5.1 illustrates the causal relationship between oil and gas resources, governance, and wealth creation. In other words, there is a great opportunity to transform oil and gas resources into wealth, poverty alleviation, human capital, institutional development, and so on, provided that basic good governance principles such as strengthened institutions, transparency, accountability, and enhanced civil society are pursued. Once economic growth and other positive results are achieved, there is a potential positive feedback loop created by turning the growth into investments and enhanced resource management and utilization. All these opportunities can be achieved and properly harnessed *only* if the good governance component is in place.

Implementing a nature–wealth–governance framework for development in Africa requires new strategies and instruments. Africa's future economic growth and development cannot be separated from the management of its natural resources, and sound environmental management cannot be separated from the broader context of politics, political systems, and governance (AfDB, 2007*a*).

The nature, wealth, and governance principles outlined above are consistent with the features, components, and instruments of other recently developed analytical frameworks for utilizing natural endowments to foster sustainable development (AfDB, 2007*a*). These include the frameworks developed from the analysis of minerals and mining by the World Bank, the United Nations Conference on Trade and Development, and the International Council on Mining and Metals (World Bank, UNCTD and ICMM, 2006) and by the Mining, Minerals and Sustainable Development project (MMSD, 2002). They are also similar to the features of new poverty reduction frameworks developed by the United Kingdom's Department for International Development (DFID, 2006*a*; 2006*b*; 2002; Pearce, 2005; UNEP and IISD, 2004).

Table 5.1 illustrates the complex challenges and opportunities that arise in the implementation of the suggested model. The main challenges, opportunities, and policy issues are listed according to interlinked main elements: oil and gas resources, governance, and wealth (corresponding to the model in Figure 5.1). The main challenges related directly to oil and gas resources include issues such as geographic distribution of oil and gas resources, the particular economic nature of the oil and gas, environmental problems, limited knowledge of the resource, lack of appropriate human capital, and inadequate contracting procedures. At the "governance level" the main challenges include weak legal and fiscal regimes, corruption, social and political conflicts, inappropriate revenue distribution mechanisms, lack of human capital, weak institutional capacity, and lack of resources to improve governance. The challenges related to the wealth include inability to attract investment into the oil and gas sector (especially the downstream industry), deficiencies in promoting economies of scale, limited cost-effective technical solutions, lack of equitable distribution of wealth, macroeconomic distortions, dependence on a few commodities (lack of diversification), price volatility (makes planning difficult), limited intra-regional trade and cooperation, and so on. This matrix of challenges can be streamlined further by distinguishing between the specific issues related to net oil-importing countries and oil- and gas-producing countries (discussed in more detail in Section 5.3).

Table 5.1 also presents potential opportunities in terms of oil and gas resources, governance, and wealth

Table 5.1: Main Challenges and Opportunities in Turning Oil and Gas Resources into Economic Growth and Policy Implications

Issues	Oil and Gas Resources	Governance	Wealth and Poverty Reduction
Challenges	• Erratic distribution of resources • Enclave nature of O&G resources • Environmental pollution (e.g. gas flaring) • Limited geological data and knowledge (resource estimate, evaluation) • Lack of skilled labor/expertise • Procedures for contract preparation, negotiation are not optimal • Refinery inefficiency	• Corruption • Social and political conflicts • Equitable growth and revenue distribution • Build capacity and invest in human resources and productive capital • Human and institutional capacity • Weak legal and fiscal regimes	• Attracting investments to the oil and gas sector • Promoting economies of scale • Limited cost-effective technology • Inequitable distribution of wealth, limited effect on poverty reduction • "Dutch disease" and macroeconomic distortions • Dependency on a single commodity (lack of diversification) • Price volatility • High oil prices for *net oil-importing Africa* • Low regional trade integration and cooperation

Opportunities	• Significant resources (more to be found) • Increased attention on exploration • Oil and gas of high quality • Learning from other countries, in Africa and elsewhere	• Transparency initiatives (EITI, APRM, PWYP, etc.) • Continued democratic development • NGOs, civil society engagement • Awareness, rights issues, etc.	• Africa is strategically located to meet the demand • Improved royalty, revenue, and tax regimes • Investment in economic diversification • Investment in human and productive capital • Link to social and physical development • Savings and stabilization (investment and future funds, etc.) • Use resources (income) for PRSPs, etc. • Promote rural development • Increased worldwide demand (including Africa)

(*cont.*)

Table 5.1: (*Continued*)

Issues	Oil and Gas Resources	Governance	Wealth and Poverty Reduction
Main Policy Issues	• Support geological surveys • Support research, innovation, and dissemination of improved technology • Support improved knowledge and contract procedure, etc.	• Support and strengthen transparency and accountability initiatives • Enhance decision-making and "local content" development • Regulation, legislation, information • Increase awareness (institutions, civil society, population)	• African Petroleum Fund (and other initiatives to address high oil prices) • Promote regional integration (trade, infrastructure, coordination, expertise) • Resource allocations and funding of investments, projects, improved procedures, etc. • Support learning processes, innovation, experimentation, diversification (workshops, training, centers of excellence, etc.) • Increase energy access, efficiency—and engage in alternative energy supplies (to fossil fuels) as a medium–long-term solution

Source: Authors.

creation. Section 5.4 explores some of these opportunities in more detail. Finally, Table 5.1 also highlights some of the main policy issues, including those that are highly relevant for regional institutions such as the African Development Bank and the African Union. These will be discussed further in Chapter 6.

5.3 Challenges

5.3.1 *Net Oil-Importing Countries in Africa*

One of the characteristics of oil and gas is high price volatility. High oil and gas prices have a dramatic impact on net oil-importing countries (NOICs) in Africa. High commodity prices have already been a motive for riots and demonstrations in many African countries (for example, in Mozambique on February 5, 2008). The main reason for the public outcry is the more than 50 percent rise in public transport costs as a consequence of the increase in the pump prices for diesel and petrol. The rise in the pump price of fuel results in an increase in the general cost of living. The scenario is even worse for poor households that depend on kerosene for lighting and cooking.

Coping with high oil prices, worsened by lack of resources to establish stabilization mechanisms and price smoothing, is one of the major challenges facing NOICs (Chapter 4). Most countries have opted to either partially or completely pass through the international oil prices, while others have introduced subsidies. All these mechanisms have proved inefficient in coping with high oil prices in Africa.

5.3.2 *Net Oil-Exporting Countries in Africa*

Managing oil wealth: Oil and gas exploitation have caused major economic, social, political, and environmental problems in some net oil-exporting countries. Oil wealth has not supported sustained economic growth and development in most countries. The major challenge is to translate oil wealth into sustained economic growth and development.

Particular challenges faced by oil-rich countries include weak governance, low accountability, low capacity for proper budgeting and accounting, and lack of transparency in the oil and gas industry. These lead to high investment uncertainty, and, in some cases, may fuel social conflicts. The Extractive Industries Transparency Initiative (EITI) offers an opportunity to address the issue of transparency.

Equitable allocation of the proceeds from oil is key to reducing the risk of social tensions in oil-rich countries. In Nigeria, for example, the federal government passes 13 percent of oil revenues to the nine oil-producing states, primarily to reduce conflict and promote local development. Although this may not fully satisfy the local stakeholders in the Niger Delta, it is a step in the right direction (see Box 5.1).

Box 5.1: MANAGING OIL WEALTH IN NIGERIA: THE CHALLENGE IN THE NIGER DELTA

The struggle for the control of oil wealth dates back to the 1960s. At stake has been the quintessential property rights issue: who owns the oil resources in the Niger Delta? Through legislative enactments, most notably the Land Use Decree of 1978, the federal government has vested the ownership of all land in the country to itself, including the minerals, ores, oil and gas resources found in them. This has been contested by local communities. There are three inter-related dimensions to the Niger Delta crisis: economic (resource control); environmental (the negative impact of oil exploitation on the environment); and social (health and human rights issues).

In recent years, several factors have contributed to the intensification of the conflict over the control of resources: the communities' increased sense of deprivation, the growing ecological damage, the lack of physical and social infrastructural facilities, and the deepening poverty and neglect in the region.

A solution to these problems must include community involvement in developing programs for the region, community empowerment, and restoration of trust between the communities, the oil companies, and the government. In practice, this calls for several initiatives: new or renewed efforts in hiring indigenes into operational, managerial, and executive positions in the oil and gas sector, within the government, and in the oil companies; award of maintenance and servicing contracts to indigenes; allocation of oil blocs to the communities; involving the communities in the design and implementation of programs for the region; and empowering the communities by providing them with financial resources for developing their own businesses and for skills development, so that they are active participants in the execution of the various programs being planned for the region.

In addition, equity could be promoted through the creation of decentralized trust funds for specific programs that address the key issues in the Niger Delta. These can include a physical infrastructure fund, a social infrastructure and training trust, an environmental repair trust, and a small-medium enterprise fund. The trust funds would be managed by corporate governance structures that include federal (state government) and community representatives. The arrangement will also enhance government–community partnerships, stimulate competition among the various funds to produce results, and generate jobs, especially for youths.

Source: Otobo (2007) and AfDB (2007*a*).

Economic distortion: The risk of the Dutch disease constitutes an important management challenge for oil-producing countries. Non-inflationary policies to prevent hyperinflation and maintain monetary credibility can minimize the risk of the Dutch disease. Moreover, it is important to promote non-oil industries or strengthen linkages that can be leveraged by the revenue from oil as a way to reduce dependence on a single commodity.

Environmental challenges: A large amount of the associated gas has been flared over the years at enormous economic and environmental costs locally and globally. Utilizing Africa's gas resources to benefit the region will be a major challenge in the quest to achieve expanded energy access in the region (AU, 2006). Currently, the use of gas is minuscule compared with its potential industrial and power use and LNG export should help eliminate all flared gas in the region.

Lack of institutional, technical, and human capacity: Traditional and new oil-producing countries usually lack adequate infrastructures to manage the oil and gas sector; in addition, the lack of technical and human capacity undermines the ability to maximize the gains from oil resources.

5.3.3 *Regional Integration*

A key economic challenge in Africa is securing reliable energy security and access for both consumption and industrial use. African countries *are* prioritizing expanded energy access. Dependence on small domestic markets in seeking to achieve this expanded energy access may not be economically viable and may thus not be the right path to pursue. Broadening national boundaries into expanded regional energy markets, driven by regionally based efficient energy production and supply structures, is central to securing affordable, reliable, and efficient energy supply. This objective is a major economic and political challenge that will face the continent in the next decades.

5.3.4 *Information as a Challenge*

Access to information is also a key challenge to sustainable use of oil and gas resources in Africa. Most African countries lack data on demand and supply by end use, price, cost, inventory movement, investment, and trade flows, on a timely and consistent basis. Effective policy design and analysis require that information should not only be available, but be reasonably accurate and distributed in a timely manner. Lack of information (mainly geological and market data) leads to weak bargaining power in contract and concession negotiations, resulting in suboptimal capture of rents from oil resources. Thus, national strategies for natural resource extraction must involve plans and funding for capacity development and generation of reliable and timely information.

5.3.5 *Inadequate Policy and Legal Framework*

In some African countries, oil and gas management is governed by rules, regulations, and other command-and-control mechanisms that stipulate what people can do and what they cannot do. The legal systems, which are either concessionary or contractual, have been developed mostly to address the rights and obligations of host governments and of private investors (Tordo, 2007). Regulations place conditions and requirements on the use of oil and gas resources, and establish sanctions—fines, fees, and other charges—for non-compliance. However, the enforcement of these rules, regulations, and mechanisms has been insufficient and ineffective. Many countries lack or have inadequate policies and legal frameworks for managing oil and gas resources. Other countries have unstable policy frameworks. Quite often, laws do not meet the requirements of international organizations in terms of transparency, accountability, and other good governance criteria. Also, lack of modern oil and gas exploitation policies is a severe drawback to the development of the sector. These include policies guiding contracts, documentation of exploitation, code of conduct and exploitation practice, training and development of local staff and community members, oil and gas research, financial guidelines, and environmental regulations. In some countries, the oil and gas sector has been driven by annual ministerial policy statements on the budget. Yet the importance of the sector in the African economy requires that a long-term planning approach for oil and gas development be adopted. Indeed, the absence of modern laws, standards, and codes governing oil and gas investment and management, and the lack of effective legal and regulatory enforcement mechanisms for such laws, standards, and codes, negatively impact the social, financial, and economic sectors. Potentially adverse effects are on the environment (pollution); accountability and revenue collection (evasion of taxes, duties, royalties, unjust enrichment, and corruption); trade (smuggling of conflict minerals, unauthorized exports of internationally sensitive materials); populations (displacements, lack of benefits for communities from resources); exports (slow and improper issuance of export permits); and revenue generation (suboptimal socioeconomic and financial returns from extractive resources).

Policies and legislation regulating the sector must be imperatively reviewed. Oil and gas policy objectives in African countries have to be formulated in the context of the following settings: the existing economic, social, and environmental policies; the nature and linkages of the oil and gas sector with other sectors; and international and regional linkages of the sector. Oil and gas policy should seek to meet the following broad objectives: establish the availability, potential, and demand of oil and gas in the countries; increase access to modern, affordable, and reliable oil and gas services as a contribution to poverty eradication; improve oil and gas governance and administration; stimulate economic development; and effectively and efficiently manage oil- and gas-related environmental impacts.

Fundamental to attracting meaningful investments and developing beneficial relationships for all stakeholders is the need to have sound legal and regulatory frameworks in which oil and gas projects are designed, negotiated, implemented, and managed. This entails having appropriate and reasonable investment guidelines, taxation regimes, environmental guidelines, labor laws, and enforcement mechanisms. In addition, in order to attract investment to the oil and gas sector, laws, regulations, and policies governing the industry should be clear, complete, transparent, accessible, flexible, and practical. Legal processes must be quick and remedies efficient and effective. Stability of fiscal contract terms is also essential. Also, a consultative process should be institutionalized to ensure periodic dialog with operators to ensure that regulations are technically feasible and cost-effective.

5.4 Opportunities

The increasing competition for African gas—due to rising international trade, the environmental advantages of gas over oil, and production declines in the largest gas markets in OECD countries—represents an opportunity for Africa. Thus, appropriate strategies must be designed for development and expanded regional use of this premium energy resource.

Africa has a unique opportunity to learn from the experiences of other countries and continents to get it right in terms of sustainable management of its oil and gas resources.

It is important to encourage the voice of civil society in monitoring and overseeing the management of oil and gas resources with the main objective of guaranteeing equitable distribution of revenues. The international transparency initiatives (EITI, APRM, and others) that promote wider involvement of the public in oil and gas management can also help promote sustainable use of the resources. Furthermore, the continuous democratization of the continent is a key opportunity for the sustainable management of oil and gas resources.

5.5 Summary and Conclusion

Oil and gas resources can be a source of growth and development if efficiently exploited and managed. To promote broad-based development and socioeconomic equity, governments must ensure that all citizens have equitable opportunities to access and use natural resources. Governments can promote pro-poor economic growth by helping to create new opportunities that boost the assets of rural communities, especially those engaged directly or indirectly in oil and gas exploration.

Experience shows that institutional and governance failures are the root causes of much underdevelopment and environmental degradation in Africa. The absence of responsive governance threatens natural resource management, economic growth, and the well-being of the population at large (AfDB, 2007*a*).

To maximize the gains from oil and gas resources, countries must pursue a number of strategies, including the following:

- Deepening current economic, social, and political reforms at the country, sub-regional, and regional levels;
- Promoting increased domestic value-added to oil and gas in Africa through expansion of downstream activities to meet both regional and export demand;
- Deepening integration efforts to create dynamic inter- and intra-regional markets for goods, including energy. This must address the issue of competitiveness, for example;
- Ensuring transparency in oil and gas exploitation and resource wealth management;
- Strengthening the fiscal and legal framework for oil and gas market operations and regulation;
- Defining roles for key stakeholders in the region, including the African Union, the AUC, UNDP, AfDB, and other development institutions, such as the World Bank, the OPEC Fund, and the EU;
- Raising awareness about the moral hazard problem associated with providing financial resources for countries affected by higher oil and gas prices. Such resources have the potential to reduce the incentives

of beneficiaries from taking the required policy measures for the emergence of a more efficient domestic energy market that helps reduce vulnerability to external oil market shocks;

- Making a long-term and credible regional and sub-regional commitment to a sustainable energy vision anchored on the emergence of efficient oil and gas markets in the next few decades; and
- Adopting regional and sub-regional strategies for developing energy security.

6

Conclusions and Recommendations

6.1 Introduction

Oil and gas resource endowments are a source of wealth and can be expected to support growth and development if well managed. This involves converting the wealth into physical and human capital, increasing and sustaining economic growth, and alleviating poverty. This report has clearly outlined that the successful pathway to fully harnessing the benefits of oil and gas resource wealth in Africa is nested in good institutions and governance, trade links, sound economic policies, and heavy investments in exploration technology, legislative framework, and human capital development.

Historically, oil and gas resource endowments in Africa have *not* yielded positive economic and development results for many countries. The experience has been more positive recently, and there are encouraging signs of better management and governance. Only history will tell whether this is a lasting change; the challenges are clearly still considerable for oil- and gas-rich Africa.

Oil and gas resources can be advantageous to economic development by providing a platform for three critical wealth elements:

1. Diversification away from dependence on agricultural primary products;
2. Investments in human capital development; and
3. Infrastructure development.

6.2 Sustainable Management of Oil and Gas Revenues

African leaders, regional institutions, and all key stakeholders have stated unequivocally that good governance is a key prerequisite for development. One of the key objectives of the African Union is to promote democratic principles and institutions, popular participation, and good governance. Furthermore, the African Union's New Partnership for African Development (NEPAD) identifies good governance as a basic requirement for peace, security, and sustainable growth and development.

Accountability and transparency in the management of public resources is critical in ensuring positive development results impact—and perhaps nowhere more so than when significant natural resource wealth such as oil and gas are present. As outlined clearly in this report, having rich oil and gas resources is not in itself a "problem" or an obstacle for development in African countries. While there are often clear linkages between resource abundance, on the one hand, and conflicts and relative underperformance, on the other hand, the real issues include fragile institutions, poor policy choices, inappropriate government practices, and inappropriate utilization of resources, especially inefficient investment of revenues.

6.2.1 *Economic and Social Measures for Maximizing the Benefits of Oil and Gas Resources*

The objective of maximizing oil and gas resource revenues and benefits for current and future generations can be broken down into four key elements, which correspond to different stages, from extraction through revenue flows to expenditure:

1. The first concerns contracts with exploration companies and appropriate fiscal regimes;
2. The second, transparency in natural resource payments and spending of the resulting revenues (including suggestions for more effective public sector financial management);
3. The third, the timing of natural resource expenditures, in particular, how this can benefit the general population and future generations; and
4. The fourth, investment and consumption decisions.

6.2.1.1 CONTRACTS AND FISCAL REGIMES

This is a crucial stage where the government has to ensure that the contracts reflect state interest and support environmental principles and sustainable extraction. Furthermore, the social responsibility of the companies has to be well expressed in the contracts in order to guarantee that communities will directly benefit from exploitation of the resources. The contracts also have to ensure the participation of local companies and investment of part of the revenues back in the country. In Africa, experiences and better practices that lead to equitable distribution of oil and gas revenues are gradually gaining momentum. In order to succeed, countries need to equip themselves with appropriate expertise, and with facts and data that can argue in favor of the state interest.

Oil and gas companies inevitably have better access to information on their specific business than African governments do. This information gap is sometimes compounded by an "agency" problem, which may occur when a deal is struck between the representative of a company and the representative of a country. Companies are in a better position to ensure that their representatives are acting in the interests of the company; whereas it is difficult for governments to ensure that their representatives are acting in the interests of the country. The information gap and the agency problem, combined, reduce the benefits that countries derive from their resources.

How can these information problems be addressed? An auction is an attractive option in a simple environment, since competition between firms will secure both efficiency—maximization of net revenues—and maximum net transfers to the government. Auctions also have the great merit of ensuring transparency, if conducted appropriately. Historically, exploration rights were attributed on the basis of public tenders where preselected companies were invited to present their proposals. These procurement procedures are slowly being abandoned in favor of auctions. In the oil and gas sector, it has almost become standard practice that exploration blocks are auctioned (for example in Angola, Chad, and Mozambique). An auction of the rights to exploration oil and gas blocks is the institutional practice that generally maximizes government revenues. The huge advantage of an auction, compared with other ways of selling rights, is that it uses competition between companies to reveal the true value of the rights. This redresses the informational disadvantage of the government and also addresses the government agency's weaknesses. A well-conducted auction vastly reduces the scope for government officials to put private gain before national interests (see AfDB, 2007*a*; Collier, 2007).

6.2.1.2 REVENUE TRANSPARENCY

In oil- and gas-rich countries, the general public perception is that the country harnesses large amounts of money from royalties, taxes, and contract signature bonuses, and that the governments mismanage such revenues or that the revenues benefit only a small part of the population. Sometimes this is, indeed, a verifiable perception, but not always. Regardless of the actual status, the perceptions can be more informed, and management can be improved significantly if the government and the companies decide to participate in public accountability and transparency initiatives.

As outlined earlier in the report, some of these initiatives include the Extractive Industries Transparency Initiative (EITI), the African Peer Review Mechanism (APRM), and Publish What You Pay (PWYP).

6.2.1.3 THE FLOW AND COMPOSITION OF EXPENDITURES

It is normally a function of governments to define the priority sector where the oil and gas revenues will be allocated. Researchers on natural resources have elaborated on the best way to use the accrued revenues and suggest the following approaches:

1. Accumulation of foreign assets;
2. Domestic investment—public or private;
3. Domestic consumption—public or private; and
4. Savings or investment funds.

The choice—or combination of choices—the government makes has to be based on sound information on resource reserves, future prices, and rates of return, as well as on the political economy of the country at the time of the decision.

6.2.1.4 ESTABLISHMENT OF STABILIZATION FUNDS

The governments or regions should establish and maintain stabilization funds using revenues from oil and gas resources. This is currently the case in Norway, São Tomé, the United States, Thailand, the Republic of Azerbaijan, and Nigeria. Such action will address the challenges of volatile prices. A nation can thus also spread the benefits of its oil and gas resource wealth over a long period and safeguard the interests of future generations (intergenerational benefits). Indeed, oil funds should be established to serve the purposes of stabilization, investment, and intergenerational equity enhancement. However, such funds should be integrated with the budget to strengthen fiscal policy coordination and public spending efficiency. They should not have the authority to spend. Financing funds should be preferred to funds with rigid rules. These funds devolve the focus of fiscal policy design and implementation to the budget and highlight the importance of non-oil balance for fiscal programming. Expenditure decisions should be reflected in non-oil fiscal indicators and in the net accumulation of assets in the oil fund, which, in some cases, may make fiscal policy more transparent (IMF, 2007). In addition, stringent mechanisms should be implemented to ensure transparency, good governance, and accountability; prevent the misuse of resources; and provide assurance that government assets are properly and prudently managed.

Channeling fiscal revenues from oil towards pro-poor investments: Some countries—such as Norway and the

Gulf (Middle Eastern) countries—have successfully used their oil wealth to stimulate growth for poverty reduction. They have done this by levying taxes on the extraction of natural resources and channeling the proceeds to poverty-reducing investments. This may include earmarking certain oil resource revenues for marginalized groups—often those living near the oil fields themselves or people from the oil-bearing communities.

Maximizing the productivity of existing natural capital through complementary public investments: Most types of natural resources require complementary investments to effectively generate growth. Many such investments are unlikely to be financed by the private sector and require public support. Examples include physical infrastructure such as irrigation facilities, which can sharply increase agricultural productivity, and transport infrastructure, which can provide access to markets and inputs.

Promoting diversification away from oil: Developing a successful modern economy based on natural resource exports is, in principle, feasible, given the right institutions and policies, as demonstrated by OECD countries such as Canada, Australia, or the Scandinavian countries. However, it is critical to use oil resources to develop a more diversified economic structure. Some policies are helpful in fostering diversification. These include establishing a conducive business environment and providing sufficient incentives to invest in non-oil sectors. A conventional measure is to use the tax system to assist the development of non-oil sectors. In addition to tax policy, there is also need for structural reforms, including financial sector and administrative reforms, to facilitate the diversification of economic activity. In many oil-dependent African economies, there is a large scope to reduce the burdens imposed by heavy regulation and an often corrupt bureaucracy, which, in addition to strengthening the financial system, would help create a more level playing field and decrease barriers to entry.

Development of alternative energy sources—Case for renewable energy: Over the past decades, conventional energy options, largely based on fossil fuels, were seen as economically more attractive than renewable energy options. However, the economic case for various forms of renewable energy is improving rapidly. According to OECD (2008), this is mainly due to three important trends: *First*, the volatile world market prices for conventional energy sources, in particular, oil, pose great risks for large parts of the world's economic and political stability, with sometimes critical effects on energy-importing developing countries. Concerns have been expressed that the rising cost of oil, which exceeded $147 per barrel in July 2008 before falling drastically in the last quarter of the year, may slow down recent economic progress in Africa and lead to tighter financial constraints. Moreover, dependence on imported fuels leaves many oil- and gas-importing countries vulnerable to disruption in supply, which may pose physical hardships and an economic burden for others. This situation has made it urgent for many countries to look for alternatives in order to make them less vulnerable to shocks in the fossil fuel markets.

Second, in the last few years, renewable energy technologies have experienced substantial improvements in cost, performance, and reliability, making them competitive today in a range of applications. Evidence shows that the cost of generating electricity from renewable energy sources has dropped significantly over the last 25 years and is expected to decrease further in the coming years. This steady fall in renewable energy prices has considerably improved the cost competitiveness of several renewable energy options, making them better placed to compete on the energy market. According to the OECD/IEA, small hydro power and biomass, are already competitive in many wholesale electricity markets, whereas in certain regions, such as in Africa, wind and geothermal energy are cheaper than conventional energy sources on the retail consumer market. Other technologies, such as solar PV, solar water heaters, and biomass, are often the most cost-effective options for providing energy services in off-grid areas in developing countries.

Third, there is increasing evidence that human activity, especially the combustion of fossil fuels, is responsible for the global warming trend which has been observed over the past decades. Also, evidence is increasingly showing that, particularly in Africa and other developing countries, climate change will drastically reduce economic growth. As the Stern (2006) report shows, the benefits of strong, early action on climate change, among other things, through the promotion of renewable energy, will outweigh the cost of dealing with the effects of climate change in the future.

Thus, despite the abundance of renewable energy resources (wind, sunshine, biomass and hydro) and the clear benefits for stimulating (pro-poor) growth, renewable energy has not yet realized its potential on the ground. The main reason is that the existing energy markets include a number of obstacles that frustrate the development of mature markets for the application of renewable energy. Some policies that will help harness the potential benefits of using renewable energy include: creating the right environment for attracting foreign direct investment (FDI); establishing a pro-poor playing field for decentralized energy technologies; developing renewable energy sector policies and measures; creating small-scale financing systems for renewable energy; and bridging the gap between development projects and small local entrepreneurs (OECD, 2008).

Improving and increasing intra-African trade in oil and gas: There is relatively little oil trade among African countries. Reducing transport costs within the continent is key to stimulating trade. Estimates indicate that shipping a car from Japan to Abidjan, Côte d'Ivoire, for example, costs $1,500, whereas shipping the same car from Addis Ababa, Ethiopia, to Abidjan costs $5,000 (ECA, 2004). In general, each additional day a shipment is in transit is equivalent to an extra 0.8 percentage point increase in applied tariffs. Africa needs to devote more resources to regional infrastructure. Such investment is also necessary to enhance domestic competition in an integrated regional market. The second key area relates to strengthening cross-border cooperation. The cost of crossing a border in Africa can be equivalent to the cost of traveling more than 1,000 miles inland, whereas in Europe the cost is equivalent to traveling 100 miles. African countries could also cooperate in a range of other areas, such as energy, water resources, research and education, environment management, and the prevention and resolution of regional conflicts. Unlike preferential trade agreements, such cooperation does not lead to trade diversion.

The third area is streamlining regulatory requirements regarding the technical specifications of oil products. The specifications tend to vary from country to country, forcing oil refineries to produce relatively small amounts of different types of products (at high cost), and reducing the potential for regional trade. There is much scope for coordinating taxation and pricing policies (at the very least, this may make part of the existing oil trade visible in government statistics). However, one of the key obstacles to larger intra-African oil trade relates to counterparty risk. Exporters from one country are simply unable or unwilling to take on the counterparty and sovereign risks for a buyer from another country. Instead, they prefer to sell on cash to a developed country trader, who then takes the risk on the foreign buyer. Structured financing techniques can help overcome credit risk problems. These techniques can go from the relatively simple (for example, one can "discount," or forfait (the seller makes a one-time arrangement with a bank to take over responsibility for collecting the accounts receivable) the payment obligation of the buyer/importer with

Afreximbank, the Cairo-based African trade finance bank set up by the African Development Bank, receiving immediate cash while Afreximbank takes on the payment risk on the buyer/importer), to the more complex (for example, certain export receivables from the importing country can be assigned for payment of the oil purchases) (Rutten, 2001).

Political reforms and good governance: Future oil management must be based on deepened reforms, expanded energy access in support of the Millennium Development Goals (MDGs), increased domestic value-added, transparency, and a long-term vision. The key international oil revenue management initiative, the EITI, is concerned with the generation of revenue and not the use of revenue. There is a need for transparency and accountability all along the supply chain and on the spending side. A lot of revenue leakage occurs during the building of infrastructure and execution of other government projects. Indeed, experience on how to manage fluctuating incomes with price volatility is needed. However, without government will, effective oil revenue management will never happen. Public dialog and local management of resources with inclusion of civil society is also crucial, as is a political and social contract for managing oil revenues, based on democratic participation and transparent economic governance. The success of such governance will depend on political stability, government legitimacy, a long policy horizon, high domestic savings and, in particular, powerful non-oil political constituencies.

6.3 Implications for the AU, the AfDB, and Other Development Partners

Successfully harnessing resource windfalls to make a significant contribution to Africa's economic growth and development is naturally a major task for governments, but also a central concern for the African Union (AU), the African Development Bank (AfDB), and all other development stakeholders. The analyses presented in this report reveal the existence of a unique opportunity for donors, multilateral financial institutions, and others to influence the role of resource wealth in addressing Africa's development challenges.

6.3.1 *Implications for and Role of the African Development Bank*

The AfDB has been involved in the extractive industries (EI) sector—including the oil and gas sub-sector—for nearly three decades. However, according to recent assessments by a commissioned task force, the contribution of the AfDB to the development of the sector is relatively modest and falls short of the sector's potential. The cumulative volume of the Bank's lending in the EI sector only represents a very small share of the Bank's total lending (3.4 percent). In addition, the Bank's activities in the sector have not been implemented within the framework of a comprehensive and coherent strategy. Furthermore, it is only quite recently that the Bank has devoted full attention to governance issues in its EI activities and has taken transparency considerations into full account in respect of its internal procurement rules and procedures (AfDB, 2007*c*). A guidance note for the EI sector has also been approved to provide operational guidance on sound and sustainable management of natural resource sectors in the Bank's lending operations, particularly in private sector operations.

Given the sector's high potential to contribute to economic and social development in Africa, particularly in resource-rich countries, the AfDB Task Force recommended that, for these countries, the Bank consider making EI development a strategic priority and increase its involvement in the sector (internal AfDB documents). To effectively and efficiently play this role, the Bank must define a clear set of development objectives for the EI sector. This is particularly important for resource-rich fragile states and post-conflict countries, where EI sector revenues could potentially play a large role in speeding up economic and social development. To this end, the Bank is developing *a comprehensive strategy* for its increased engagement in the sector. The strategy will cover all aspects of EI sector development (lending as well as non-lending activities), including investment, capacity building, governance, contracts, negotiations, legal and regulatory frameworks, as well as environmental and social impacts. Moreover, to improve corporate effectiveness, the strategy will enhance internal synergies between Bank departments and sectors and mainstream governance, transparency, and accountability concerns into sector interventions (infrastructure, energy, and so on).

The following specific recommendations constitute the backbone of the full strategy and will thus (with refinements and relevant additions) form the basis for AfDB support and engagement in the oil and gas sector:

1. ***Mainstream EI sector issues into Bank operations***. The Bank should ensure that policy dialog with RMCs, and the relevant Bank diagnostic and related analytical work (such as Governance Profiles) address EI sector issues in a systematic and comprehensive manner.
2. ***Enhance internal capacity and intra-Bank coordination***. To increase its involvement in the EI sector, the Bank must build appropriate expertise while promoting greater coordination between the different units of the Bank involved in EI sector activities.
3. ***Complement financial and technical assistance activities in the EI sector with policy dialog and non-lending activities***, including economic and sector work, enhancement of knowledge and dissemination of information, advocacy, and promotional initiatives.
4. ***Work in partnership with other development partners involved in the EI sector***. This will help ensure greater synergies and complementarities in development partners' interventions in the EI sector.

5. ***Strengthen legal and regulatory frameworks***. Fundamental to attracting meaningful investments and developing beneficial relationships for all stakeholders is the need to have sound legal and regulatory frameworks in which EI projects are designed, negotiated, implemented, and managed. This entails having appropriate and reasonable investment guidelines, taxation regimes, environmental guidelines, labor laws, and enforcement mechanisms. The Bank should thus provide technical support to RMCs to develop such frameworks.
6. ***Build human, institutional, and technical capacity in relevant policy ministries and agencies***. There is a need to strengthen capacity in the EI sector in most resource-rich RMCs. This is particularly urgent for fragile states and post-conflict countries, as well as for countries that have only recently become producers of extractive resources. Strengthening the capacity of these countries will help them guard against the mistakes of others, improve chances that the extractive resources will be properly managed and that revenues can contribute to the sustainable development of the country. The Bank can play a major role in this regard by providing support to strengthen oversight bodies, such as boards of concessions; anti-corruption agencies; ministries of finance, trade, mines, and petroleum and environment; and other agencies involved in the management of the EI

sector. Within the framework of technical assistance, the Bank should collaborate with other partners to provide support for contract and concession negotiations, improve information relevant to the EI sector, and enhance the capacity of RMCs to regulate and manage the environmental and social impacts of EI projects. In particular, the Bank can play a crucial role in defining and promoting "best practices."

7. ***Improve governance and promote transparency and accountability in the development and management of EI sector operations.*** Extractive resources can contribute to sustainable development only if revenues derived from their exploitation are properly managed and used primarily to support development priorities. Promoting good governance, transparency, and accountability in the management of EI revenue should be at the center of the Bank's involvement in the sector. The Bank should pay greater attention to governance and revenue management issues in its EI sector operations. For instance, the Bank is already helping countries to improve the quality of public financial management (PFM). Encouraging transparency in the management and use of EI revenue will complement the Bank's efforts to enhance the quality of PFM in RMCs.
8. ***Adapt the nature of the Bank's interventions to the governance situation*** in RMCs and tailor

interventions to the specific requirements of each country. In particular, where the governance (macro or sectoral) situation is weak, the Bank should help strengthen governance while supporting the expansion of the EI sector, and, at the same time, adopting a rights-based approach.[1] This implies that the Bank should systematically assess the governance context and risks, as well as the social impact and poverty reduction potential, in all EI operations, especially in areas such as management and use of EI revenues.

9. ***Support EI sector initiatives***. In line with the proposed role of the Bank regarding governance and transparency in the EI sector, the Bank should support broad initiatives in the EI sector. In particular, it should encourage RMCs to adopt EITI principles, including helping them develop capacity in accounting for EI revenues, and help them publish information about these revenues in a transparent, consistent, and useful way. In this regard, the Bank *is* providing both financial and technical assistance to support implementation of the EITI and collaborating closely with selected countries and, where appropriate, other development partners, throughout the implementation process.

[1] A rights-based approach to development sets the achievement of human rights as an objective of development. It invokes the international apparatus of human rights accountability in support of development action and is concerned not just with civil and political rights (the right to a trial, not to be tortured), but also with economic, social, and cultural rights (the right to food, housing, a job).

10. ***Catalyze private investment in the sector.*** While addressing the challenges associated with the EI sector (governance, legal and regulatory frameworks, capacity, and so on), the Bank should proactively support private investment and be more engaged in the sector. In other words, although the Bank's interventions in individual projects may sometimes be limited in terms of size, it should generally play an important catalytic role by leveraging large amounts of funds from other sources.
11. ***Build information and geological data.*** The lack of adequate geological data is one of the major constraints to the development of the EI sector in many RMCs. The Bank should provide support to RMCs to improve their knowledge as well as information capacity about their extractive resource wealth, including offshore resources. In this regard, the Bank should provide funding for geological surveys and join other partners to support the establishment of an African geological databank.[2] Through the creation of such a databank, African countries will gain access to more accurate information about their extractive resources and the associated costs of exploration, thereby considerably strengthening their negotiating positions in the award of concessional agreements and contracts. The databank will also improve the capacity

[2] For example, the Africa-Europa Georesource Observation System (AEGOS) or a Regional Oil and Gas Data Center, as proposed by the World Bank.

to provide technical, economic, and environmental information that can facilitate the assessment of EI development projects by potential investors.

12. ***Establish a rapid response advisory facility for RMCs, in collaboration with other development partners***, capable of providing legal and non-legal advice and other support related to the EI sector on short notice. Such assistance is being provided within the framework of the Bank's African Legal Support Facility established by the Bank.[3] The work of the Facility would contribute to capacity building in RMCs over time, especially in the area of contract negotiation. In particular, the Bank can also act as a resource institution for the new oil- and gas-producing countries (or countries with prospects for future production).
13. ***Support for alternative energy resources***. The Bank also needs to position itself to support projects on alternative energy sources, which range from renewable sources to low carbon emission alternative energy technologies such as clean coal technology. The development of these alternative solutions could help eradicate continental poverty, create new jobs, ensure the security of energy supply, and combat global warming, thus contributing to global justice, security, and peace. The Bank can provide this support through African initiatives on infrastructure

[3] This Facility is particularly useful for fragile and post-conflict states, where human capacity is scarce.

for alternative energy sources, including NEPAD's Short-Term Action Plan (STAP), which focuses on priority physical investments in energy, transport, water and sanitation, and information and communications technology (ICT), complemented by capacity building, policy, regulatory, and institutional measures aimed at improving the efficiency of the existing infrastructure on the continent.

14. ***The Bank will play a key role in operationalizing the envisaged African Petroleum Fund,*** in cooperation with the African Union. The Fund will mobilize the international community and African nations to provide resources to assist net oil importers that are experiencing balance-of-payment problems.
15. ***Finally, the Bank needs to analyze and position itself with regard to the increasing Asia–Africa trade partnerships, with*** focus on empowering oil-producing countries to understand and address different options and implications in pursuing loans, investments, and business cooperation where oil and gas resources are at stake.

6.3.2 *Implications for the African Union*

The African Union plays an important political role in financial resource mobilization and implementation in the oil and gas sector. The political backing for financing the energy future of the region lies with the AU mandate and its institutions. Promoting new financing options for sustainable energy systems needs strong and credible political backing of the highest order. The AU should use all available international and regional fora to sensitize potential donors and other stakeholders to support the proposed African Petroleum Fund Initiative. As promising as this initiative may be, the initial costs, over $1 billion at least, may lead some stakeholders to adopt a wait-and-see attitude. Without a conscientious effort by the AU to push the Fund Initiative, Africa's energy and development future—through expanded energy access at affordable prices and oil and gas wealth sustainability—may remain an elusive goal.

The African Peer Review Mechanism (APRM) provides an avenue for monitoring governance in the sector as part of the overall peer review of member countries under NEPAD. The AU should also provide strong support to strengthen weak political, economic, and institutional frameworks for proactive sub-regional and regional approaches to economic and energy development.

6.3.3 *Implications for Other Development Stakeholders*

The role of other development partners include: assistance in financial resource mobilization, capacity building in the oil and gas sector, including training in the use of innovative oil and gas financing options, energy infrastructure management, governance issues in the industry, and policy strengthening at the country, sub-regional, and regional levels. Such development partners include the United Nations, United Nations Environmental Program (UNEP), Organization for Petroleum Exporting Countries (OPEC), OPEC Fund, International Monetary Fund (IMF), World Bank, European Union, United States Agency for International Development (USAID), the UK's Department for International Development (DFID), Agence Française de Développement (AFD), Swedish International Development Cooperation Agency (SIDA), and other bilateral and multilateral institutions. Nevertheless, as clearly outlined in this report, the main responsibility for transforming oil and gas wealth into other "capital" and for addressing high oil prices lies with the national leadership and governments in Africa.

6.4 Recommendations

The analysis in this report suggests a number of recommendations at the national, regional, and continental levels.

6.4.1 *At the Continent Level*

At the continental level, it is important that pan-African institutions, mainly the AU and the AfDB, become more engaged in the management of oil and gas resources in Africa. These resources represent a crucial element and contribution to the economic and social development of the continent.

At this level, a solid understanding of the multi-faceted nature of Africa's oil and gas challenges and opportunities—as outlined in detail in Chapter 5—is called for. Indeed, in a multi-ethnic and often polarized and impoverished society, the management of oil and gas resources will remain an important factor for social and political stability.

It is likewise very important that the continental bodies, together with other international donors and stakeholders, provide new or prospecting oil and gas producers with technical and financial assistance. This will help these producers negotiate fair and pro-development contracts where environmental, social, and revenue distribution is an integral part of the management of oil and gas.

The AfDB and the AU are urged to pool human, political, and financial resources to help both oil-producing and oil-importing countries improve exploitation of resources and seize the various opportunities at hand.

6.4.2 *At the Regional Level*

Regional economic blocs such as Southern African Development Community (SADC), Common Market for Eastern and Southern Africa (COMESA), and Economic Community of West African States (ECOWAS) can likewise play a key supporting and coordinating role, and are urged to become more proactive in the oil and gas sector, notably through the following actions:

1. Promote regional integration in the exploitation of oil and gas.
2. Promote regional infrastructures (such as oil and gas pipelines) for sustainable exploitation of oil and gas.
3. Promote regional economies of scale in the sector, especially with focus on the downstream industry (for example, refineries).
4. Encourage countries to adhere to principles of good governance and transparency initiatives for revenue management.
5. Promote regional sharing of experiences.
6. Promote intra-regional trade in the oil and gas sector.
7. Support and promote the African Petroleum Fund.
8. Establish regional fora for sharing experiences on oil and gas issues, especially those related to contract negotiations and technology transfer.
9. Promote the use of regional experts in the sector.

6.4.3 *At the Country Level*

At the country level, it is important to distinguish between oil producers and net oil importers. The following recommendations are proposed for the following specific categories of countries.

Oil-Producing Countries: The experience of African countries reveals that natural resource endowments, especially in non-renewable resources, come with opportunities, but also with substantial risks. The key challenge for harnessing natural resource opportunities is to design the right strategic policies in a context that supports fiscal prudence and minimizes macroeconomic distortions. This should be backed by adequate institutional capacity and national and local-level participation in oil and gas management.

Getting the policies wrong, disregarding their sequencing and alignment with the rest of the economy, or ignoring absorptive capacity and good governance issues may transform a natural resource boom into a curse that could effectively stall economic growth, worsen the poverty situation, and become a recipe for social and political instability.

While recognizing that countries have specific backgrounds and circumstances, the following recommendations can nevertheless apply to most of the oil-producing countries:

1. Develop knowledge of verified and potential oil and gas resources (geological data and resource estimation, etc.). Establish an oil and gas information database.
2. Manage oil and gas resources sustainably and value high standards of environmental protection, including rehabilitation.

3. Secure optimal rents (royalties, taxes, shares, and other revenues), based on sound, fair negotiations (auctions recommended) and "best practices" for contracts.
4. Engage in and promote good governance, accountability, and transparency in oil and gas revenue management. This also entails engagement in initiatives that can reduce corruption (EITI and APRM) and secure added transparency and engagement of civil society.
5. Promote revenue distribution that reaches local people in the oil fields. Make public awareness on the revenue distribution a way to accommodate public speculations. Design and implement pro-poor policies to ensure equitable distribution of benefits.
6. Engage in social policies that promote harmony and well-being in the communities affected by oil and gas production.
7. Promote economic diversification as a way to avoid commodity dependence (investment in the future).
8. Promote policies and ensure fiscal regimes that prevent market distortions—such as the "Dutch disease."
9. Strengthen the human, technical, and institutional capacities of the relevant institutions.
10. Design policies that encourage investment in non-oil sectors.

11. Manage income for optimal intergenerational benefits, for example, by establishing sovereign wealth funds or other savings and investment funds.
12. Seek advice from regional and continental stakeholders and supporters (such as the AfDB and the AU), especially in contract negotiations.
13. Support regional and continental initiatives that seek economic, social, and political improvements in oil, gas, and energy-related initiatives—encompassing regional integration and the African Petroleum Fund initiative.
14. Support regional infrastructure for the sustainable exploitation of oil and gas.
15. Support regional economies of scale in the downstream industry.
16. Support intra-regional trade of oil and gas.
17. Share countries' good and bad experiences with regional and continental stakeholders.

Net Oil-Importing Countries: The following recommendations apply to net oil-importing countries:

1. Identify and recognize the direct and indirect impacts of high oil prices.
2. Engage in alternative fuel sources such as biodiesel.
3. Find adequate mechanisms for mitigating/smoothening the effects of high oil prices. Such mechanisms may include

- Price-based policies (comprising full passing on of price increases and granting targeted subsidies to some vulnerable end-users);
- Policies to reduce the cost of supplies (hedging, economies of scale, security of stocks);
- Quantity-based policies (reduce demand and oil consumption, for example, by increasing use of public transport);
- Diversification into non-petroleum sources of energy, from fossil fuels to renewable power sources; and
- Increasing domestic supply.

4. Support the African Petroleum Fund and, when it is fully established, subscribe and contribute to its operation.
5. Identify potential downstream industries where the country has a competitive advantage (for example, a refinery in an accessible location).

Bibliography

Abeysinghe, T. (2001), "Estimation of Direct and Indirect Impacts of Oil Price on Growth," Economic Letters, 73.

AfDB (2005), *African Development Report 2005: Private Sector Development in Africa*. Oxford University Press, Oxford.

——(2007*a*), *African Development Report 2007*. Oxford University Press, Oxford.

——(2007*b*), (AfDB Memorandum on EITI)—Support to Implementation of EITI in Resource Rich Regional Member Countries. African Development Bank/African Development Fund, Tunis.

——(2007*c*), (AfDB Memorandum on APRM)—Implementation update to Bank Group support to APRM. African Development Bank/African Development Fund, Tunis.

African Energy (2008), Website database. www.africa-energy.com.

Al-Attar, A., and Alomair, O. (2005), "Evaluation of Upstream Petroleum Agreements and Exploration and Production Costs," *OPEC Review*, Vol. 29, No. 4, pp. 243–66, December 2005.

Alley, P., C. Bermann, L. Danielson, H. Feldt, S. Mahalingam, A. Nadal, C. Nair, S. Nguiffo and S. Siakor (2007), To have and have not: Resource Governance in Africa in the 21st Century. A memorandum of the Heinrich Boll Foundation.

APRM (2007), African Peer Review Mechanism (APRM) Homepage. http://www.aprm.org.za.

AU (2006), First African Union Conference of Ministers responsible for Hydrocarbons (oil and gas) Concept Paper. Addis Ababa, Ethiopia.

——(2007), Conference of Ministers of Industry, 1st extraordinary session. Midrand, South Africa.

Auty, R. (2001), *Resource Abundance and Economic Development*. Oxford University Press, Oxford.

——(2004), "Economic and Political Reform of Distorted Oil-Exporting Economies. Presentation at The Earth Institute, Columbia University.

Ayadi, O. F. (2005), "Oil Price Fluctuations and the Nigerian Economy," *OPEC Review* Vol. 29, No. 3, pp. 199–217.

——Amitava Chatterjee, and C. Pat Obi (2000), "A Vector Autoregressive Analysis of an Oil-Dependent Emerging Economy—Nigeria," *OPEC Review*, Blackwell Publishing.

Bacon, R. (2001), Petroleum Taxes: Trends in Fuel Taxes (and Subsidies) and the Implications, World Bank note no. 240. www.worldbank.org/html/fpd/notes/.

Barsky, R. B., and L. Kilian (2004), "Oil and the Macroeconomy Since the 1970s," *Journal of Economic Perspectives*, Vol. 18, No. 4, pp. 115–34.

Bouakez, H., and D. Vencatachellum (2007), The Impact of High Oil Prices on African Economies, African Economic Conference, 15–16 November, Addis Ababa, Ethiopia.

BP (2006), *BP Statistical Review of World Energy 2006*. London, UK.

——(2007), *BP Statistical Review of World Energy 2007*. London, UK.

BP (2008), *BP Statistical Review of World Energy 2008*. London, UK.

Bruno, M., and J. Sachs (1982), "Energy and Resource Allocation: A Dynamic Model of the Dutch Disease," *Review of Economic Studies*, Vol. 49, pp. 845–59.

Castells, M. (2000), *End of Millenium*. Blackwell Publishing, UK.

Chmaytelli, M. (2005), "Sudan to double its refinery capacity." http://www.busrep.co.za.

Chow, M. (2007), Seattle Biotech Legacy Foundation in conjunction with the Institute for Children's Environmental Health, Seattle, US.

Christianson, D. (2006), "Sudan: Getting In at Ground Level," *Business in Africa*, at www.businessinafrica.net/energy_in_africa/997130.htm.

CNGMO (2007), "Natural gas market dynamics: Record storage levels and mild weather curb prices in 2006," *Canadian Natural Gas Market Outlook—Winter 2006–0;* Canadian Centre for Energy Information.

Coady, David, and David Newhouse (2006), "Ghana: Evaluating the fiscal and social costs of increases in domestic fuel prices," in A. Coudouel, A. Dani, and S. Paternostro, eds., *Analyzing the Distributional Impacts of Reforms: Operational experience in implementing poverty and social impact analysis*. World Bank, Washington.

——Moataz El-Said, Robert Gillingham, Kangni Kpodar, Paulo Medas, and David Newhouse (2006), "The Magnitude and Distribution of Fuel Subsidies: Evidence from Bolivia, Ghana, Jordan, Mali, and Sri Lanka," IMF Working Paper WP06/247, Washington DC.

Collier, P., and B. Goderis (2007), "Commodity Prices and Growth: Reconciling a Conundrum. Department of Economics," University of Oxford.

—— (2007), "Managing Commodity Booms: Lessons of International Experience." CSAE, Department of Economics, University of Oxford.

Council for Geoscience (CGS) and Mintek (2007), GIS Database. MESU, Randburg, South Africa.

Daniel, J. A. (2001), "Hedging Government Oil Price Risk," IMF Working Paper WP/01/185, IMF, Washington DC.

Davidson, O., and Y. Sokona (2002), "Energy and Sustainable Development: Key Issues for Africa (Abstract)," *Bulletin Africa*, No. 15. http://unpan1.un.org/intradoc/groups/public/documents/IDEP/UNPAN008520.pdf.

DFID (Department for International Development) (2002), *Better livelihoods for poor people: the role of land policy*. London.

—— (2006*a*), *Eliminating world poverty: Making governance work for the poor.* DFID White Paper, London.

—— (2006*b*), *Eliminating world poverty: Helping to build states that work for poor people*. DFID White Paper, London.

Dohner, R. S. (1981), "Energy Prices, Economic Activity and Inflation: Survey of Issues and Results,". In Knut Anton Mork (ed.), *Energy Prices, Inflation and Economic Activity*. Ballinger, Cambridge, Mass.

Dudine, Paolo, James John, Mark Lewis, Luzmaria Monasi, Helaway Tadesse, and Joerg Zeuner (2006), "Weathering the Storm So Far: The Impact of the 2003–05 Oil Shock on Low-Income Countries," IMF Working Paper 06/171, Washington.

EDF (Environmental Defence Fund) (2007), "Scenes from the Chad-Cameroon Oil and Pipeline Project." http://www.edf.org/article.cfm?contentID = 2135.

EIA (Energy Information Administration) (2004), International Energy Annual Accounts (2000–2004).

EIA (2006), *International Energy Outlook 2007*, EIA.

—— (2007), *International Energy Outlook 2007*, EIA.

—— (2008). http://www.eia.doe.gov.

EITI (2007), Extractive Industries Transparency Initiative (EITI) homepage. http://www.eitransparency.org.

Engineering News (2007), "New Mozambique oil refinery approved." http://www.engineeringnews.co.za/article.php?a_id=115115.

Equator Principles (2007), Equator Principles homepage. http://www.equator-principles.com.

ESMAP (Energy Sector Management Assistance Programme) (2004), "Energy and Environmental Health: A Literature Review and Recommendations"; ESMAP Technical Paper 050, Washington DC.

—— (Energy Sector Management Assistance Programme) (2005*a*), ESMAP Annual Report, ESMAP, Washington DC.

—— (Energy Sector Management Assistance Programme) (2005*b*), "Power Sector Reform in Africa: Assessing Impact on Poor People," Report 306/05, ESMAP, Washington DC.

—— (Energy Sector Management Assistance Programme) (2006), "Coping With Higher Fuel Prices" Report 323/06, Washington DC.

Gavin, M. (1990), "Structural Adjustment to a Terms-of-Trade Disturbance: The Role of Relative Prices," *Journal of International Economics*, Vol. 28, pp. 217–43.

—— (1992), "Income Effects of Adjustment to a Terms of Trade Disturbance and the Demand for Adjustment Finance," *Journal of Development Economics*, Vol. 37, pp. 127–53.

Gazprom eyes Nigerian reserves: http://www.fin24.co.za/articles/default/display_article.aspx?ArticleId=1518–1783_2247100.

Hamilton, James D. (2000), "What is an Oil Shock?," NBER Working Paper 7755, June.

——(2005), "Oil and the Macroeconomy," forthcoming in *The New Palgrave Dictionary of Economics*, ed. by S. Durlauf and L. Blume (London: Macmillan, 2005, 1st edn). Available at: http://dss.ucsd.edu/~jhamilto/JDH_palgrave_oil.pdf.

——(2006), "Oil and the Macroeconomy," forthcoming in *The New Palgrave Dictionary of Economics*, ed. by S. Durlauf and L. Blume (London: Macmillan, 2006, 2nd edn). Available at: http://dss.ucsd.edu/~jhamilto/JDH_palgrave_oil.pdf.

Heinrich Boll Foundation (2007), Expert Round Table: Resource Governance in Africa in the 21st Century, Berlin, 26–28 March.

Hibou, B. (1999), "The Social Capital of the State as an Agent of Deception," in J. Bayart, S. Ellis, and B. Hibou (eds.), *The Criminalization of the State in Africa*. James Currey, Oxford.

Hossain, S. M. (2003), "Taxation and pricing of petroleum products in developing countries: A framework for analysis with application to Nigeria," IMF Working Paper WP/03/42, IMF, Washington DC.

Hunt, B. (2005), "Oil price shocks: Can they account for the Stagflation in the 1970s?," IMF Working Paper WP/05/215, IMF, Washington DC.

——P. Isard, and D. Laxton (2001), "The macroeconomic effects of higher oil prices," IMF Working Paper WP/01/14, IMF, Washington DC.

IEA (International Energy Agency) (2004), "An Analysis of the Impact of High Oil Prices on the Global Economy." http://www.iea.org/textbase/papers/2004/high_oil_prices.pdf.

—— (2008), *Energy Balances of Non-OECD Countries-2008*, IEA, Paris.

IMF (2000), *World Economic Outlook: Asset Prices and the Business Cycle*. International Monetary Fund, Washington DC.

—— (2005), *World Economic Outlook: Building Institutions*. International Monetary Fund, Washington DC.

—— (2006), *World Economic Outlook*. International Monetary Fund, Washington DC.

—— (2007), *Regional and Economic Outlook: Sub-Saharan Africa*, World Economic and Financial Surveys. IMF, Washington DC.

—— (2009), IMF Commodity Prices Database, January 2009.

International Petroleum Monthly Review (2006), "International petroleum," http://findarticles.com/p/articles/mi_m2744/is_ 3_2006/ai_n16359904.

James, L. W. (2007), Oil Price History and Analysis, WTRG's homepage.

Jourdan, P. (2006), "Regional Strategies—the Case for a Resource Based Spatial Development Programme," Presentation delivered at the US–Africa Infrastructure conference, 28 September.

—— (2007), "The Role of the Extractive Industry in Fragile States in Africa," Presentation at the African Development Bank Annual Meeting.

Kaufmann, D., A. Kraay, and M. Mastruzzi (2007), Governance Matters VI: Governance Indicators for 1996–2006. World Bank Policy Research, World Bank, Washington DC.

Kilian, L., A. Rebucci, and N. Spatafora (2007), "Oil Shocks and External Balances," IMF Working Paper WP/07/110, IMF, Washington DC.

Kisambira, E. (2007), "Tullow oil refinery plan *East African Business Week* – http://www.busiweek.com/index.php?option=com_content&task=view&id=4490&Itemid=9.

Kpodar, K. (2006), "Distributional Effects of Oil Price Changes on Household Expenditures: Evidence from Mali," IMF Working Paper WP/06/91, IMF, Washington DC.

Krichene, N. (2006), "World Crude Oil Markets: Monetary Policy and the Recent Oil Shock," IMF Working Paper WP/06/62, IMF, Washington DC.

Lee, Kiseok, Shawn Ni, and Ronald A. Ratti (1995), "Oil Shocks and the Macroeconomy: The Role of Price Variability," *Energy Journal*, Vol. 16, pp. 39–56.

Loungani, P. (1986), "Oil Price Shocks and the Dispersion Hypothesis", *Review of Economics and Statistics*, 68.

Malaquias, A. (2005), "Thirsty powers: the United States, China, and Africa's energy resources." Unpublished report, St. Lawrence University, Canton, NY.

Manin, B., A. Przeworski, and S. C. Stokes (1999), Introduction, in: *Democracy, Accountability and Representation*, ed. Adam Przeworski, Susan C. Stokes, and Bernard Manin. Cambridge University Press, Cambridge.

Market Research (2008), Market Research.com. http://www.marketresearch.com.

Mining, Minerals and Sustainable Development (MMSD) (2002), *Breaking new ground: The Mining, Minerals and Sustainable Development Report*. IIED, Earthscan, London.

Moore, M., and G. Teskey (2006), The CAR Framework: Capacity, Accountability, Responsiveness, What do these terms mean, individually and collectively?, A discussion note for Department for International Development Governance and Conflict Advisers' Retreat, 14–17 November.

Mork, K. (1989), "Oil and the Macroeconomy, When Prices Go Up and Down: An Extension of Hamilton's Results," *Journal of Political Economy*, Vol. 97, No. 51.

—— (1994), "Business Cycle and Oil Market," *The Energy Journal*, Vol. 15.

Nageria, "Oil and Gas Downstream Sector (Refining and Marketing) Report—2005." http://www.marketresearch.com/product/display.asp?productid=1125089&g=1.

NGV, http://alternative-energy.everything-fine.com/?p=145.

Nicholson, K., B. O'Laughlin, A., Fransisco, and V. Nhate (2003), "Fueled tax in Mozambique," PRSP Monitoring and Synthesis Project, Poverty and Social Impact Analysis Briefing Paper.

NNPC (2008), Nigerian National Petroleum Corporation. Website information—http://www.nnpcgroup.com.

Obadina, T. (2000), "Harnessing abundant gas reserves: Nigeria alone could provide the power needs of all West Africa." *Africa Recovery online*, A UN publication.

ODI (2006), *Meeting the Challenges of the Resource Curse: International Experiences in Managing the Risks and Realizing the Opportunities of Non-Renewable Natural Resources Revenues*, Overseas Development Institute, London.

OECD (2008), *Natural Resources and Pro-Poor Growth: The Economics and Politics*, DAC Guidelines and Reference Series. OECD, Paris.

Oil and Gas Journal (2007), Online data available at: http://www.ogj.com/index.cfm.

Onlinenigeria.com (2005), "Sahara Petroleum to build oil refinery in Nigeria," http://nm.onlinenigeria.com/templates/?a = 3779&z = 2.

Ostry, J., and C. Reinhart (1992), "Private Saving and Terms of Trade Shocks: Evidence from Developing Countries," Staff Papers, International Monetary Fund, Vol. 9, pp. 495–517.

Otobo, E. E. (2007), "Managing of oil resources wealth in Nigeria—The challenges in the Niger Delta," short paper prepared for AfDB (in relation to the *African Development Report*, 2007).

Otto, J. M. and J. Cordes (2004), *Regulation of Mineral Enterprises: A Global Perspective on Economics, Law and Policy*. Rocky Mountain Law Foundation, Westminster, Colorado.

Pearce, D. (2005), "Managing environmental wealth for poverty reduction, poverty and environment partnership," *MDG 7 Initiative Economics,* 31 August.

Ploeg, F. van der (2007), "Africa and National Resources," Background Paper for AfDB, *African Development Report.*

PWYP (2007), Publish What You Pay homepage (PWYP). http://www.publishwhatyoupay.org.

Research Division (2007), *The Impact of High Oil Prices on African Economies,* Working Paper Series (WPS) No. 93, December 2007, African Development Bank Group, Tunis.

Reuters News (2008), "Gazprom eyes Nigerian reserves," REUTERS. http://www.fin24.com/articles/default/display_article.aspx?ArticleId=1518–1783_2247100.

Ribot, J. C. (2006), "Choose Democracy: Environmentalists' Socio-political Responsibility," *Global Environmental Change,* Vol. 16 pp. 115–19.

Rogers, C. D. (2007), "Sustainable Development, Poverty and Livelihood." UNECA (www.uneca.com).

Ross, M. I. (1999), "The Political Economy of the Resource Curse," *World Politics*, Vol. 51, pp. 297–322.

Rutten, L. (2001), Using new financial instruments for improving oil trade across Africa: Opportunities and limits, African high-level meeting on energy and sustainable development, UNEP/UNDESA, Nairobi, January 10–13, 2001.

Sapa–Agence France-Presse (2007), "Sudan Oil Output to Reach 600,000 bpd," *Business in Africa*, April 3, at www.businessinafrica.net/news/north_africa/.

Schultz, R. (2007), "Africa's Energy Crisis Worsens: Viable Clean Energy Alternatives Are Imperative." http://www.americanprogress.org/issues/2007/07/energy_crisis.html.

Semboja, H.H.H. (1994), "The Effects of Energy Taxes on the Kenyan Economy," *Energy Economics*, Vol. 3, pp. 205–15.

Sigonney, P. (2006), Oil supply prospects and market dynamics; Total, IFRI Paris, Social Impact Analysis Briefing Paper.

Stern, N. (2006), *The Economics of Climate Change. The Stern Review*. Cambridge University Press, Cambridge, United Kingdom.

Stevens, P. (2003), "Resource Impact—Curse or Blessing?: A literature survey," Centre for Energy, Petroleum and Mineral Law and Policy, University of Dundee, Dundee.

Terry, D. (2007), "Ethanol from Biomass: America's 21st Century Transportation Fuel." www.ethanol-gec.org/information/DT_Ethanol_from_Biomass_Update_8-31-07.ppt., Governors' Ethanol Coalition.

The Institute for Children's Environmental Health. http://www.iceh.org/pdfs/SBLF/s.pdf.

Thibodeaux, R. (2005), "Anger Rises in Oil-Rich Chad as Funds Don't Aid the Poor," *The Boston Globe*, at www.boston.com/news/world/articles/2006/04/30/anger_rises_in_oil_rich_chad_as_funds_dont_aid_the_poor.

Tordo, S. (2007), "Fiscal Systems for Hydrocarbons: Design Issues," World Bank Working Papers No. 123, World Bank, Washington DC.

U.S. Department of Energy (2006), U.S. Department of Energy Strategic Plan. http://www.energy.gov.

——(2007*a*), Energy Information Administration, "Sudan," *Country Analysis Brief*, at www.eia.doe.gov/emeu/cabs/Sudan/Background.html.

——(2007*b*), Energy Information Administration, "Chad and Cameroon," *Country Analysis Brief*, at www.eia.doe.gov/emeu/cabs/Chad_Cameroon/ *Background.html.*

UNDP (2006*a*), *Human Development Report 2006: Beyond Scarcity: Power; Poverty and the Global Water Crisis*. United Nations Development Programme, New York.

——(2006*b*), http:// hdr.undp.org / hdr2006 / statistics/ documents/hdi2004.pdf.

UNECA and AfDB (2007), "The 2007 Big Table: Summary Document, Managing Africa's Natural Resources for Growth and Poverty Reduction," Addis Ababa.

United Nations Environment Programme (UNEP) and International Institute for Sustainable Development (IISD) (2004), "Exploring the Links: Human well-being, poverty and ecosystem services," Nairobi, UNEP.

United States Agency for International Development (USAID) with Center for International Forestry Research, Winrock International, World Resources Institute and

International Resources Group (2002), "Nature, wealth, and power: Emerging best practices for revitalizing rural Africa," USAID, Washington DC.

Whateley, M. K. G., and P. K. Harvey (1994), *Mineral Resources Evaluation II: Methods and Case Histories*, Geological Society Special Publication No. 79. The Geological Society, London.

World Bank (2000), World Development Indicators, World Bank, Washington DC. Available at: http://publications.worldbank.org.

——(2006*a*), "Engaging with Fragile States: An IEG Review of World Bank Support to Low-Income Countries Under Stress," World Bank, Washington DC.

——(2006*b*), "World Development Report 2006," World Bank, Washington DC.

——(2006*c*), "Where is the Wealth of Nations? Measuring Capital for the 21st Century," International Bank for Reconstruction and Development/World Bank, Washington DC.

——(2006*d*), "World Development Report 2006: Equity and Development," World Bank, Washington DC.

——(2006*e*), World Bank Data for Gross National Income, World Bank, Washington DC.

——(2007*a*), World Development Indicators, 2007, World Bank, Washington DC.

——(2007*b*), "World Development Report, 2007: Income Data," World Bank, Washington DC.

World Bank, United Nations Conference on Trade and Development, and International Council on Mining and Metals (2006), "The Analytical Framework: The Challenge of Mineral Wealth: Using resource endowments to foster sustainable development," World Bank, Washington DC.

WTO (2007), International Trade Statistics, WTO, Geneva.

Zissis, C. (2007), "Chad's Oil Troubles," Council on Foreign Relations, updated, at www.cfr.org/publication/10532/chads_oil_troubles.html.